BATTLEFIELD OF THE FUTURE

21st Century Warfare Issues

Barry R. Schneider

Lawrence E. Grinter

Revised Edition

September 1998

The Air War College Studies in National Security was established as a forum for research on topics that influence the national security of the United States. Copies of No. 3 in this series are available from the Air War College, 325 Chennault Circle, Maxwell Air Force Base, Alabama 36112-6427. The fax number is (334) 953-7934; the telephone number is (334) 953-2103/DSN 493-2103/7074.

Air War College
Studies in National Security No. 3

Air University Press
Maxwell Air Force Base, Alabama 36112-6428

Library of Congress Cataloging-in-Publication Data

Battlefield of the future : 21st century warfare issues / [edited] by Barry R. Schneider and Lawrence E. Grinter—Rev. ed.
p. cm.—(Air War College studies in national security : no. 3)
1. Military art and science—Forecasting. 2. Twenty-first century. I. Schneider, Barry R. II. Grinter, Lawrence E. III. Series.
U104.B38 1998
355.02'01'12—dc21 98-38913
CIP

ISBN 1-58566-061-2

First Printing September 1995
Second Printing (Revised Edition) September 1998
Third Printing July 2001
Fourth Printing August 2002
Fifth Printing September 2004
Sixth Printing May 2007

Disclaimer

The views expressed in this publication are those of the authors and are not the official policies or positions of the United States Department of Defense or the United States Government.

Air University Press
131 West Shumacher Avenue
Maxwell AFB, AL 36112-5962
http://aupress.maxwell.af.mil

To Keegie (Schneider)
who so fully supported one of the editors in
the preparation of this book, and to Gertrude E.
and George H. Schneider and Constance L. and
Linton E. Grinter who lovingly and unselfishly
gave us a twentieth-century vantage point
to peer into the twenty-first century

Battlefield of the Future
21st Century Warfare Issues

Contents

Illustrations

Acknowledgments

We, the editors, wish to thank the authors for their timely responses to our requests for quality work and for meeting deadlines. At the Air War College, we wish to thank Dean Ronald Kurth and Col Theodore Hailes, associate dean for their championing of the Air War College Studies in National Security series. Thanks also are due to Col Robert Hinds, our department chairman, who gave us his support and encouragement when we began work on this project. We also appreciate Maj Gen Peter D. Robinson, Dr Alexander S. Cochran, Dr Robert Wendzel, Dr David Sorenson, Dr Dan Hughes, Dr Bill Martel, Col Joseph Englebrecht, and Dr George Stein, all Air War College faculty, who have taken their time to review chapters written by others. Outside reviewers who also contributed valuable suggestions include: Mr David Kay (SAIC), Col John Ellen, USAF, retired (SAIC), Col Phillip Gardiner, USAF, retired (SAIC), Dr Robert Joseph, director of the NDU Counterproliferation Center; and Phillip E. Lacombe, managing director of the Aerospace Education Foundation. Help was also provided by the Institute for National Security Studies, USAFA, that supported part of Dr Schneider's research on this project. We are also grateful to MSgt Milton Turner for running interference on the budgetary paperwork, and to Modeyither Jones and Linda Jenkins for their administrative assistance that helped us turn this book out in a timely fashion. At the Air University Press, we wish to thank Dr Elizabeth Bradley, director; Tom Mackin; Thomas Lobenstein; Rebecca McLeod; Steven Garst; and other AU Press editors that gave us valuable suggestions for improving the manuscript in a timely fashion.

Barry R. Schneider & Lawrence E. Grinter

Introduction

This is a book about strategy and war fighting in the midst of a revolution in military affairs as the world moves into the twenty-first century. Its 11 essays examine topics such as military operations against a well-armed rogue state or NASTI (NBC-arming sponsor of terrorism and intervention) state; the potential of parallel warfare strategy for different kinds of states; the revolutionary potential of information warfare; the lethal possibilities of biological warfare; and the elements of an ongoing revolution in military affairs (RMA).

The book's purpose is to focus attention on the operational problems, enemy strategies, and threats that will confront US national security decision makers in the twenty-first century. The participating authors are either professional military officers or civilian professionals who specialize in national security issues. Two of the architects of the US air campaign in the 1991 Gulf War have contributed essays that discuss the evolving utility of airpower to achieve decisive results and the lessons that might portend for the future of warfare.

In "Principles of War on the Battlefield of the Future," which sets the tone for the book, Dr. Barry Schneider examines how traditional principles of war may have to be reassessed in light of a proliferation of weapons of mass destruction (WMD) among third world states. Regarding the principle of "mass," traditional theory dictated that forces be massed for an offensive breakthrough. But Schneider argues that, against an enemy armed with WMD, *dispersal* of one's forces may, in fact, be more prudent, and fighting by means of "disengaged combat" prior to a decisive strike may be necessary. This requires high coordination and "superior targeting and damage assessment intelligence, combined with superior high-tech weapons."

Still, the United States and its allies would not likely be able to dominate a future battlefield even with advanced conventional arms if they did not have close-in air bases to operate from and thereby to achieve air dominance over the battle space. Therefore, while it sounds good, striking from outside the enemy's range is not a real option for long if the enemy is mounting a ground campaign that is closing in on vital areas. Local air, sea, and ground power will be needed to contain the adversary forces and roll them back. This means local air bases and seaports must be available and protected.

Yet, the US and allied armed forces, in confronting a Saddam Hussein with nuclear, biological, or chemical (NBC) warfare capabilities might be well advised to forego massing forces (which present lucrative targets to the enemy's WMD) in favor of maneuver, dispersion, speed, mobility, range, and deception. How to protect fixed installations such as ports and airfields is a dilemma. Furthermore, Schneider tells us, the principle of maintaining the "offensive" may have to be supplemented "with a combination of potent defenses to avoid lethal enemy [WMD] counterstrikes."

In twenty-first-century warfare, theater missile defenses (TMD) are likely to be essential, especially against future rogue regimes possessing nuclear, biological, or chemical warheads and ballistic missile delivery systems. These could pose a threat against US and allied forces, ports, airfields, naval convoys, and cities within range that only effective multilayered TMD may be able to handle.

When facing a WMD-armed adversary, it will be even more important than in the past to preserve "unity of command" via effective command, control, and communications during a conflict. Moreover, it is highly likely that, in an era of "information warfare," both sides will attempt and may be able to disrupt and destroy each other's command and control systems.

Regarding the principle of clear, obtainable "objectives," Schneider argues that war with a nuclear-armed terrorist state "must either be a short victorious war that starts with the neutralization or destruction of the enemy's WMD, or one fought for limited objectives and prosecuted with deep respect for the power of the adversary's mass destruction capabilities." This would require a revolution in the way US regional "war-fighting" commanders in chief prepare for major regional conflicts (MRC). It might be difficult for US decision makers to "sell" such a strategy to an American public, given our penchant for quick, decisive victories.

Finally, "security" as a principle of war demands exceptionally good intelligence. In the future, it will be especially important to identify those states acquiring WMD and missile capabilities and to gauge their locations and numbers from the outset of a conflict. This will be difficult because rogue proliferator states greatly complicate accurate intelligence on their paths to acquiring WMD by pursuing multiple clandestine paths to such capabilities. These paths include the use of underground secret facilities, camouflage, disinformation, dispersal, cheating on Non-Proliferation Treaty requirements, purchasing "dual use" technologies, and other means to disguise their programs and hide their facilities. The discovery after the Gulf War of how far Iraq had progressed in acquiring nuclear, biological, chemical, and ballistic missile capabilities was a wake-up call to the world community.

In summary, the dangers of confronting an adversary with weapons of mass destruction or the capacity for strategic information warfare may prompt a very different thinking about the traditional "principles of war." Preparing for such an eventuality will require some major changes in C^3I (command, control, communications, and intelligence), military doctrine, operational strategy, acquisition, equipment, logistics, coalition building, coalition warfare, and war termination. Some changes may even be necessary in foreign policy regarding the kinds of commitments that US vulnerabilities, capabilities, and interests will permit in a more proliferated world or one where the enemy has utilized the technologies and methods of information warfare.

Chapter 1

Principles of War for the Battlefield of the Future

Barry R. Schneider

The United States would have fought its wars of the past half century far differently had Hitler, Mussolini, Tojo, Kim Il Sung, Mao Tse-tung, Ho Chi Minh, Manuel Noriega, and Saddam Hussein possessed nuclear weapons at the time.

A world of nuclear-armed states will require the United States and its allies to revise force structures, strategy and doctrine, intelligence capabilities, command and control procedures, and logistics for major regional conflict scenarios. A proliferated world of potential adversaries equipped with weapons of mass destruction and the means of delivering them will require the US military to implement a "revolution in military affairs," one that may require significant departures from current US strategy, operational policies, and military capabilities.[1]

Clearly, US force planning and conflict preparation have not yet taken into account a "Saddam Hussein with nukes" to use Les Aspin's phrase when he announced the US Defense Counterproliferation Initiative. The *Bottom Up Review*, conducted by the Clinton Administration under then-Secretary of Defense Aspin, did not assume the United States would confront an adversary armed with weapons of mass destruction in either of the two nearly simultaneous major regional conflicts (MRCs) that US forces are supposed to be able to fight and win. Yet, it is clear that radical and hostile states such as Iraq and Iran are probably just a few short years away from having a nuclear weapons capability and North Korea may already possess one. All three are presently credited with biological and chemical weapons capabilities.

Implications for Military Strategy

So how do you fight a NBC-armed sponsor of terrorism and intervention (NASTI) on the battlefield, if war breaks out? Do the old principles of war work in this kind of conflict? And just what are those principles which have guided US and allied forces in past wars? In the United States, even young ROTC students are taught the elements of war, summed up by the acronym MOSSCOMES:

M - Mass
O - Offensive
S - Surprise
S - Security
C - Command Unity
O - Objective
M - Maneuver
E - Economy of Force
S - Simplicity

Seven of these principles were extracted from the works of British major general J. F. C. Fuller, who provided them for the instruction of the British Army in World War I.[2] They were then republished in a 1921 US Army training regulation and have been passed on in Air Force, Army, and Joint doctrine and professional military education publications since.[3]

Some of General Fuller's ideas may be applied without modification to future war against hostile radical adversaries armed with weapons of mass destruction. Other principles of war have to be modified to reflect changes in technology or situation. For example, WMD in enemy hands suggests that future commanders modify the way they apply the war-fighting principles of mass, maneuver, command unity, and taking the offensive initiative in combat. New technology provides new stealthy means of achieving surprise, and end-of-war residual enemy WMD capability may very well alter allied approaches to security and war termination ends and means.

Further, there are some additional principles of war that Fuller did not address that deserve attention in an era marked by the proliferation of weapons of mass destruction. These include the advantages to be gained by simultaneity and depth of attack, effective force-projection logistics, information dominance, and precision targeting.

What is new and what is constant in this brave new proliferated world? Let us look first at the principle of "mass" in warfare.

The Principle of Mass in Warfare

The principle of mass suggests the wisdom of concentrating superior combat power at the decisive place and time in military operations in order to achieve decisive results.[4] This massing of resources directed at key enemy vulnerabilities helps one's own forces to retain the initiative and makes it possible, together with the proper application of other principles of war, for outnumbered forces to achieve break-throughs and decisive war, campaign, and battle results.

For example, Mao Tse-tung in his guerrilla war strategy emphasized the importance of achieving local superiority in battle even though one's own forces were greatly outnumbered overall in the conflict across all major theaters. His tactics when engaging the enemy called for ten against one, even if outnumbered ten to one at the strategic level. In Mao's strategy, proper choice of the time, place, and ratio of engaged forces could shift victory from the hands of larger-but-more diffused enemy forces, to those of less numerous-but-more highly concentrated forces that achieved greater mass at the points of contact.[5]

When J. F. C. Fuller wrote his treatise on the principles of war in World War I, mass was strongly correlated with numbers of ground troops concentrated in a given location against enemy ground forces in close proximity. Today, such massed units would be vulnerable to a different type of mass derived from weapons of mass destruction and precision guided munitions delivered by missiles, aircraft, or superguns.[6] This gives a new meaning to "local superiority."

Ideally, US forces can catch regional opponents in a paradigm shift, where the adversaries may adhere to the older notions of mass—that is massing of their armies. US forces can substitute the application of massed firepower for massed troops. In such a competition, massed allied firepower could put to flight or destroy massed enemy units.

The increased lethality of conventional weapons has led to progressively greater dispersion of forces in the field with each passing era. For example, the density of troops deployed in the battle zone has decreased from an average of 4,790 troops per square kilometer in the Napoleonic Wars to just 2.34 troops per square kilometer in the 1991 Persian Gulf War. WMD threats will accelerate a historical trend toward a progressively emptier battlefield.[7]

As military technology has improved over time, firepower has increased and the size of units directing it and trying to avoid it has decreased.

As one analyst observes:

> The logical end point of such developments (advanced conventional arms and WMD) is the replacement of the notion of concentration of mass with one emphasizing concentration of fire. Increasingly, modern armies of the future should achieve breakthroughs and victory without resorting to large masses of troops directed at vulnerable points. Instead, the combination of rapidly firing systems, precision weapons of long range, and advanced command and control systems will allow widely dispersed forces to focus their fire on specific points.[8]

Indeed, in combat with an adversary armed with WMD, one corollary to Fuller's dictum on "mass" is that dispersing one's own forces can make enemy WMD less cost-effective. A second corollary is that massed allied firepower needs to be directed first to destroying or degrading enemy WMD at the inception of combat to permit the later massing of one's own general purpose forces for combat in the war-termination phase of the conflict.

Just as in the American Civil War and World War I, when massed offenses were slaughtered by heavily concentrated defensive firepower, the future possession of WMD in enemy hands should discourage the use massing of allied troops until after the opponents WMD are silenced or neutralized.[9] If it looks like disabling early strikes cannot neutralize enemy WMD in a projected conflict, perhaps such an adversary should not be engaged in the first place, provided that is an option (i.e., if the war has not yet begun and if one's homeland and forces are not already engaged).

Unfortunately, in some inherently unstable situations, if the adversary were to strike first with weapons of mass destruction, he might achieve victory, at least temporarily, in a regional conflict. If the adversary is vulnerable to an allied preemptive strike, he would have an incentive to use his WMD first. Such a perceived "use or lose" situation is inherently unstable and unpredictable, especially in a crisis or escalating conflict.

Some analysts even suggest that the development of very advanced conventional armaments, combined with new strategy and organization of forces, can be a "revolution in military affairs," making the massing of troops impractical and dangerous. Thus, one of three courses of action may be adopted by the allied commander when faced with a NASTI armed force:

- Desert Storm II: Proceed as if the threat did not exist, except to rely upon escalation dominance to deter the adversary from escalating to WMD use in the conflict.
- Dispersed Storm: Adopt many measures to protect the allied force, such as disinformation, extended dispersal of units, downsizing of units, constant mobility, passive defenses, and active defenses while still engaging in traditional forms of warfare, relying also on escalation dominance to preserve intrawar deterrence of enemy WMD use.
- Remote Engagement: Adopt a mode of "disengaged combat," where allied forces conduct their military operations at a substantial remove from their enemies.[10]

The first approach is the same approach that the United States and its coalition took with regard to possible Iraqi use of its biological warfare (BW), chemical warfare (CW), and Scud assets in the 1991 Gulf War. In this conflict, despite the vulnerability of allied forces and capitals, the allies used counterforce strikes and active and passive defenses to protect against Iraqi air and Scud attacks and used escalation dominance to deter the possible Iraqi use of available BW and CW assets. This combination might work again in the future if the adversary is similarly outclassed in the air, and where the preponderance of high-tech weapons is held by the allies. Nevertheless, it is a risky strategy that might backfire with huge downside results.

The second approach is where forces similar to those sent to the Gulf War are given far more protection, by much improved air defenses, missile defenses, and passive defenses. The regional CINC would also reduce the number of lucrative theater targets available to the enemy by an extensive dispersal of his own forces and logistical units and by very pronounced use of mobility to increase enemy uncertainty concerning the location of key allied forces.

The third approach is where the main allied force stays outside of enemy range and attempts to pick off his WMD and destroy his massed forces by air, missile, and special forces attacks before sending the bulk of the expeditionary force to engage him in the endgame. In remote engagements, the allied force would attempt to outrange the adversary and degrade his capability before closing and attempting to finish the conflict on allied terms.

In the future, friendly forces may be well advised to avoid, where possible, close massed engagements with heavily armed enemy forces. Instead, they likely should adopt the Dispersed Storm or Remote Engagement postures as a mode of operations out of respect for the possible consequences of an enemy WMD strike, particularly if the adversary develops a capability well beyond that achieved by Iraq in 1991.

There are trade-offs in adopting the Dispersed Storm mode of operations. On the one hand, failure to mass one's own troops can make them more vulnerable to enemy conventional attacks. Moreover, it would be difficult to conduct normal conventional operations in a dispersed mode. On the other hand, one would run less risk of having main force units obliterated by enemy WMD strikes in this mode. The tradeoffs of adopting the Remote Engagement mode of operations, when facing an enemy with WMD, has received less discussion, and deserves to be considered first.

Some would argue that if the United States is faced with such a formidable opponent, the allies probably should first attempt to outrange them, dealing punishment from a distance while staying out of harm's way. In the words of former heavyweight boxing champion, Muhammed Ali, US and allied forces should "float like a butterfly, sting like a bee." On the other hand, getting bogged down in massed armor and artillery duels, providing mass targets to enemy advanced weapons, is a route to heavy casualties and possible defeat.

If military forces follow the strategy of disengaged combat, the battle front may be hard to find. Indeed, in such remote engagement warfare, it may not exist. The initial stages of combat might find two heavily armed rival forces, both dispersed, striking at each other from a distance, each attempting to secure an advantage by locating and striking the other's key units and assets, while simultaneously trying to stay out of harm's way from the massive and precise capabilities of the other.

If remote engagement were adopted as a strategy, then only after sufficient damage has been inflicted on the adversary via disengaged combat, would an attempt be made to close and force a capitulation. If the adversary's weapons of mass destruction have been eliminated with high confidence, this war-termination phase of conflict might resemble more traditional forms of combat. The opening scenarios of remote combat would require great standoff capabilities, the spreading and hiding of forces, intensive intelligence, attrition of enemy advanced capabilities, effective active and passive defensive measures, and extensive coordination of fire from many diverse points to the highest priority targets on the other side.

In such conflicts, each of the armed services would need to be tightly coordinated with the others. Regional CINCs would need complete connectivity to theater forces under their command while likely having to operate from highly mobile and hard-to-target command posts.

This suggests the need for superior generalship, superior targeting and battle damage assessment intelligence, combined with superior high-tech weapons. "Using the accuracy of advanced sensors and precision weapons, US forces may be able to jockey just out of the range of enemy artillery, tanks, and battlefield missiles, picking them off in turn."[11]

This kind of remote engagement conflict would require changes in US strategy, doctrine, training, and organization. Regional CINCs, in charge of fighting major regional conflicts, would have to be schooled in a different kind of war fighting from that pursued in the 1991 Gulf War, the 1964–74 Vietnam War, the 1950–53 Korean War, or World War II. Preliminary extensive war gaming, in-the-field exercises, and operational planning for the new type of warfare would be mandatory for later success in the region of combat.

An enemy with WMD or very advanced conventional capabilities obviously poses severe dangers to choke points, ports of entry, regional air bases, and naval convoys. For example, aircraft carriers and their surrounding task forces might be very vulnerable to an adversary armed with nuclear or biological weapons. These floating airfields, capable of carrying up to 100 aircraft and holding a military population of 5,000 to 6,000, represent highly lucrative targets and may be inappropriate in the future for confronting such a very heavily armed regional foe capable of obliterating or sinking them.

The US Navy in future combat against a "Saddam Hussein with nukes" may be forced to operate from more numerous, smaller, less expensive and more dispersed platforms, emphasizing ballistic and cruise missiles rather than naval aircraft as theater strike weapons. These might be augmented by longer-range, air-refueled, naval fighter-bombers launched from carriers outside the range of enemy aircraft or missiles that carried the threat of WMD bombardment and obliteration. How far the US Navy needs to go in these directions will be determined partly by how successful it is in developing fleet defenses against ballistic and cruise missiles.

The US Army, likewise, may be forced to move away from strong reliance on heavy tanks and armored personnel carriers that fight close to enemy forces. Rather, Army units may be required to hit and locate the enemy at much greater ranges, at least in the earlier phases of battle, rather than close and attempt to destroy the NASTI enemy with heavy mechanized forces before his WMD capabilities have been neutralized. As one defense analyst observes, "such armored forces are designed to fight a war that US commanders should attempt to avoid, not bring about."[12]

US Air Force officials have become convinced that massed bomber attacks are less productive than a few stealthy bombers firing or dropping precision munitions at targets from a standoff mode. A few low-observable aircraft are now able to penetrate enemy defenses with very few losses and inflict, via increased accuracy, greater damage than whole air armadas previously could inflict using less accurate bombs and missiles.

As one Air Force analyst notes, with the revolution in accuracy, "it no longer took hundreds of bombers dropping thousands of bombs or even tens of bombers dropping scores of bombs to destroy a single target. Now, one aircraft often delivering only one weapon, could destroy one target."[13]

A third element of "mass" to be considered in combat with very heavily armed opponents is the need for whole-unit reinforcements. Armies, divisions, naval task forces, or air bases brought under NBC missile attack may suffer such wholesale losses in such short time periods that they may entirely cease to function as cohesive military units. In such horrific circumstances, front-line units may need to be replaced by entire units of similar capability and numbers, perhaps under new commanders due to the massive and traumatic nature of the losses suffered from WMD bombardment. Nuclear detonations, lethal nerve gas attacks, or clouds of deadly biological agents could annihilate entire defense sectors and open large gaps in friendly forces that could be filled only with fresh units that retained their cohesion and command, control, and communications linkages.

Thus, when confronting a NASTI, or even a hostile state possessing very advanced conventional arms, it appears wise to rethink the advisability of massing one's own troops. Consider, for example, how different the outcomes of warfare might have been in the past half century if US forces and those of the Allies had to consider German nuclear strikes against the Normandy beachhead, Italian biological weapons at Anzio, or Japanese nerve gas blanketing US invasion forces at Iwo Jima.

Faced with WMD bombardment, would the allies have been able to hold the Pusan beachhead or successfully mount the

Inchon invasion during the Korean War? Indeed, would the US nuclear threat communicated to Beijing via the Indian government have been credible if the People's Republic of China also had possessed nuclear warheads and long-range aircraft in 1953? In the 1990–91 Gulf War, how would things have been different if Iraq had possessed even a few nuclear weapons and had been prepared to use them prior to the allied ground offensive while coalition troops were massing in Saudi Arabia?

The Principle of Maneuver in Warfare

Perhaps far greater emphasis will have to be placed on maneuver, the second "M" in J.F.C. Fuller's principles of war, rather than on the first "M," mass. Inherent in maneuver is the idea that mobility enhances both offensive and defensive capabilities as well as one's ability to achieve a viable deterrent and escalation superiority in both peace and war.

Coupled with the need for maneuver is the concept of dispersion. Armies in modern times are increasingly mobile and dispersed due to increases in battlefield lethality and other technical changes. Moving and spreading out gives the adversary less probability of targeting success and less of a target to hit. Prudence would advise spreading friendly forces even more in the future to expose fewer of them to any single WMD attack.

On the other hand, this need to disperse forces can greatly hinder conventional combat capability. An army dispersed will have less capability for achieving local superiority and breakthroughs against its opponents armed forces and less opportunity for battle and war termination until the main weapons of the enemy are silenced.

The need to simultaneously guard against vulnerability to WMD attack and to conduct a conventional campaign will impose contradictory pressures on regional CINCs planning future campaigns. Such dual concerns might prevent quick, decisive engagements in the future that are based on the 1991 Gulf War model. Instead, future armies may be forced to fight more at the low-intensity warfare level or to engage in prolonged conventional wars of attrition while avoiding presenting the enemy with the opportunity for a knockout blow delivered by their WMD.

Victorious armies facing NASTIs may be more preoccupied with active defense, passive defenses, mobility, dispersion, and concealment than with conventional offensive actions that could get them annihilated. Indeed, the lethality of the future battle area may be so great that a new vision of defensive deployment is required while simultaneously adding new urgency to the locating, targeting, and destroying of enemy launchers and storage compounds for enemy weapons of mass destruction and the adversary's very advanced conventional weapons.

The Principle of Offensive Initiative in Warfare

One of the principles of war found in US military doctrine is the necessity to "seize, retain, and exploit the initiative" in combat.[14] Maintaining the offensive initiative in warfare is important to victory, and also helps avoid defeat. An enemy on his heels is seldom an enemy at your throat. There is still some truth to the old adage that the best defense is a good offense. A good offense that keeps the adversary busy defending his own forces and homeland robs him of some of the potential to carry the fight to yours.

Unfortunately, offensive operations under attack by enemy WMD, or the threat of such an attack, can be difficult to execute. US Army operations during its Combined Arms in a Nuclear/Chemical Environment (CANE) exercises have shown that enemy WMD very much hindered "Blue" forces' offensive success. As one report summarized, "during offensive operations, it was noted that:

- attacks and engagements lasted longer;
- fewer enemy forces were killed;
- friendly forces suffered more casualties;
- friendly forces fired fewer rounds at the enemy;
- fratricide increased;
- terrain was used less effectively for cover and concealment."[15]

Unfortunately, as two Army analysts point out, "the introduction of NBC weapons on the battlefield by an opponent gives him the initiative."[16] Such actions, or even the threat of WMD strikes, place allied forces somewhat on the defensive and give the initiative to the opponent, since allied commanders and units are forced to take fewer risks in exposing themselves to such lethality.

Enemy use of WMD can create residual radioactive, chemical, or biological contamination of the battle area, hindering allied ability to act for hours, days, or even weeks after their use. Protective clothing, exhaustive decontamination procedures, extensive vaccination programs, administration of antidotes, and the caution borne of fear in an anthrax, highly toxic chemical, or radioactive environment can easily degrade the offensive performance and mind-set of allied forces subject to WMD bombardment. Maneuver may also be limited in battle space so contaminated.

US Army war games suggest that enemy WMD can negatively impact allied efforts to maintain the initiative, maneuver through the battlefield, synchronize forces, and project power into certain highly dangerous and contaminated areas.[17] Moreover, while conducting offensive operations, allied forces faced with WMD threats will need to operate under a defensive shield to survive and succeed. Thus, in future wars against enemies armed with weapons of mass destruction, in contrast to General Fuller's day, it will be important to supplement offensive strikes to disarm the adversary's WMD with a combination of potent defenses to avoid lethal enemy preemptions or counterstrikes to degrade the threat.

In the classic case, when dealing with a Saddam Hussein with WMD, the US military commander is faced with a dual need. First, he would like to neutralize both the enemy leadership and his WMD potential. This means the prosecution of counter-leader targeting coupled with an all-out bombardment of likely enemy WMD capabilities and production facilities. If this opening phase of the conflict is not totally successful, the allied operations should be prepared to shift dramatically from the offensive to the defensive mode, or take enormous risks that whole sectors of the allied forces might be destroyed if not dispersed into a defensive mode.

Col John Warden, one of the air architects of the allied victory in the 1991 Gulf War, postulates that future war will feature parallel strikes aimed at all the key facets of an adversary's state and force, that, if struck nearly simultaneously, will inflict strategic paralysis and quick defeat on the adversary. Airpower, he argues, is the instrument of choice for such "parallel war."

Such simultaneous, parallel strikes are a fine example of the value of retaining the offensive initiative in warfare, and the paralysis such strikes inflicted on Iraq in 1991 shows their value in keeping an adversary from taking the offensive himself. Simultaneous, parallel, in-depth attacks throughout the battle space is likely to remain as part of US military doctrine into the foreseeable future. For example, the US Army's "Force XXI Operations" study states:

> Future American operations will induce massive systemic shock to an enemy. These operations will be meant to force the loss or deny the enemy any opportunity to take the initiative.[18]

Similarly, US air doctrine emphasizes the use of new technologies such as stealth aircraft, stealthy cruise missiles, and precision guidance to give the advantages of surprise and offensive initiative to their possessor since these weapons are difficult to detect and allow airpower to go where it wishes without major losses in pursuit of strategic or tactical targets.[19] Indeed, "aerospace power can quickly concentrate on or above any point on the earth's surface. Aerospace power can exploit the principles of mass and maneuver simultaneously to a far greater extent than surface forces."[20]

However, unless the initial offensive in such hyperwar and parallel war renders inoperable the enemy's ability to strike back with weapons of mass destruction or with his most capable advanced conventional weaponry, then the conflict may feature a parallel war air blitzkrieg coupled with the pullback and dispersal of allied ground and naval forces to provide less inviting targets to possible massive enemy counterattacks spearheaded by WMD targeted on US and allied power projection forces in the region.

If total allied dominance of weapons of mass destruction is not achieved, the endgame of a conflict will be extremely risky.

Will the enemy escalate at the end or will he be deterred from launching NBC fusillades as his regime goes under? Will he use some WMD and threaten more use still in an attempt to achieve a better end-war settlement?

Or should allied forces keep out of range until such enemy WMD can be destroyed or until the enemy leadership is killed or replaced? If this is not possible, what then? It is possible that the better part of valor might be to accept a compromise peace that leaves the adversary regime and his military in place rather than demanding total surrender as required of Nazi Germany or Tojo's Japan in 1945. If this option is rejected, the allied side risks massive casualties, perhaps numbering in the millions, before victory could be achieved against a regional foe so heavily armed.

As in the 1991 Gulf War, the location of the enemy leadership and his weapons of mass destruction may be unknown. There will be a temptation at the inception of any such conflict to target the enemy leader or leaders to create disorganization and a regime change. However, the closer such counter-leader strike attempts come to success without accomplishing the task, the greater the possibility that the enemy regime will counter with desperate measures that might include launching a nuclear, biological, or chemical weapons attack, even if they face a clearly superior allied nuclear force that enjoys escalation dominance.

How do you achieve victory or a measure of victory in regional combat with such an enemy, and how do you, at the same time, limit the damage inflicted on allied forces and allies in the region of operations? Further, how do you limit damage to the continental United States and allied countries during such regional conflicts?

Until effective US and allied theater or strategic defenses are developed and deployed in the regions where foes developing or deploying WMD are located, efforts to counter such threats will have to rely upon deterrence of the adversary or on allied conventional offensive capabilities.

While it would be the very rare contingency when the United States or allied states could successfully identify, locate, target, and destroy the force of a hostile radical state on the verge of using WMD against the American homeland, US and allied forces in the region, or allied countries, there may be a few opportunities where allied intelligence can pinpoint such preparations and strike a blow to disarm such an adversary with high confidence.

Nor is it wise to use all the military potential the United States possesses, since the use of US nuclear arms to strike the enemy WMD targets would likely entail too many political, economic, diplomatic, legal, and moral negatives.[21]

In some cases, this imperative to use conventional weapons only, would make it impossible to disarm an adversary arming itself with WMD since conventional weapons may not be capable of:

- destroying deeply buried and hardened bunkers containing WMD assets;
- area targeting of widely dispersed but "soft" mobile enemy WMD assets;
- burning enemy biological weapons ingredients that were otherwise likely to be spread across the region if impacted by conventional bombing.

For these and a number of other reasons, reliance on conventional offenses alone to end the WMD threat would be unwise, because the penetration of allied defense by even a single enemy nuclear, biological, or chemical warhead might be lethal across a wide area. Theater missile defenses are also needed.

Only the combination of offensive suppression strikes coupled with defensive interception capabilities could provide any possibility of the regional "astrodome" protection needed against such unforgiving weapons, where even a single enemy warhead "leaker" through the defenses could devastate a port, base, airfield, naval convoy, massed army, or population center.

What makes the damage limitation enterprise even thinkable, once war has begun, is that the enemy may possess only a half dozen or so of such weapons at the time of a conflict, few enough so that it is possible for an allied offense-defense combination to neutralize the threat.

The Principle of Unity of Command in Warfare

Another principle of war laid out by Gen J. F. C. Fuller is that of the requirement for unity of command. Maintaining good command, control, communications, and intelligence (C^3I) could become much easier in future MRCs as a result of the ongoing revolution in information technology available to allied commanders. This information revolution will provide more information earlier, and in far greater detail about the opponent's capabilities, locations, and activities than known in previous wars.

Moreover, such a communications revolution will lead to flatter organization structures and to greater force-wide awareness of allied and enemy dispositions in real time. This will enhance the control of central commanders while, at the same time, permitting wider dispersal of friendly forces. The US Army's "Force XXI Operations" report states that

> advances in information management and distribution will facilitate the horizontal integration of the battlefield functions and aid commanders in tailoring forces and arranging them on land.... Units, key nodes, and leaders will be more widely dispersed leading to the continuation of the empty battlefield phenomenon.[22]

The challenge to effective command, control, and communications in a major regional conflict could be immense. If the adversary has the capability of decapitating the US or allied military commands, of decapitating regional allied governments, of targeting the US National Command Authority, or of "leveling the playing field" by knocking out most allied communications with a high-altitude nuclear explosion emitting a destructive electromagnetic pulse (EMP), it could destroy the unity of command of the allied forces in the region.

If the regional adversary was at a severe disadvantage in NBC weapons, he might still make effective use of his limited capability by atmospheric nuclear bursts of EMP that could play havoc with allied telecommunications, navigation, radar, aircraft, missiles, automated guns, APCs, tanks, trucks, and any microchips or electrical circuits that are not protected against EMP.

The enemy WMD threat might even extend beyond the theater of war to the capitals of allied countries, including even Washington, D.C. It may be possible that the adversary has aircraft or missiles capable of reaching such capitals. Even if this was not technically possible, it is conceivable that nuclear, biological, or chemical weapons could be delivered against such cities by unconventional means via saboteurs smuggling them in the allied countries and detonating them or threatening to do so to achieve favorable diplomatic concessions at the end of the conflict.

Unfortunately, most allied capitals are highly vulnerable to WMD threats. For example, Washington, D.C., has long been a vulnerable target and will remain so in the foreseeable future.[23] A clandestine nuclear detonation in the city would likely doom the US president, the vice president, Cabinet members, the Joint Chiefs of Staff, and members of Congress who were there at the time. The chaos that such an attack would cause would be difficult to overstate. One of the more difficult questions to answer in the hours after such a NASTI decapitation attack would be "who is in charge here?"

This chaos would be compounded if the headquarters housing the US regional CINC and his staff also were to suffer a similar decapitation strike at the same time. It is possible that the national leadership and the regional military forces of the United States would be plunged into chaos for sometime.

The threat of communication disruption and command disablement in conflicts with NASTIs leads to several conclusions regarding the preservation of unity of command in such conflicts:

- Command unity may have to give way to subcommand dispersal under a preset unified contingency plan;
- Military units may need to be more autonomous and dependent on prewar planning of operations;
- Unit commanders will need simpler, less frequent updates from central headquarters;
- Alternative commanders in mobile and hardened command posts will be needed for all regional and supporting CINCs, with trained backups in reserve several layers deep, ready to assume command if and when the CINCs are targeted, killed, or isolated from their forces;

- Military forces may have to be guided and organized similarly to distributed computer networks, with greater autonomy, independence of action, and ability to operate independent of central command while still following command guidelines.

The Principle of Clear Obtainable Objectives in Warfare

Wars, like chess matches, are generally characterized by opening moves, both offensive and defensive, by a middle game exchange, and by a decisive endgame.[24] Central and theater commanders should begin each phase of the conflict with the desired end in mind, with each phase designed to move the situation forward toward the goal. The United States Army Field Manual FM 100-5 states that commanders ought to "direct every military operation towards a clearly defined, decisive, and attainable objective."

In the Persian Gulf, President Bush defined the US and allied objective simply as the freeing of Kuwait from Iraqi occupation and the establishment of agreed borders between Iraq and Kuwait. Once beaten in the field of battle, the regime of Saddam Hussein was allowed to remain in power, although restrictions were placed upon Iraqi military units, UN inspectors were sent into Iraq to locate its nuclear, biological, and chemical weapons as well as its ballistic missiles for the purpose of destroying them. Iraq was prohibited from most international trade or exports, and was especially limited from profiting from oil exports until it was deemed to be in full compliance with peace terms negotiated at the end of the Gulf War.

President Bush's decision to stop the fighting when he did was controversial. Many thought he should have directed US and allied forces to proceed on to Baghdad when he had the Iraqi military on the run and in chaos, continuing the conflict so long as Saddam Hussein and his cabinet controlled the Iraqi government and military forces.

It has been argued that President Bush's decision was made in line with the principle of war that says to direct every military operation towards a clearly defined, decisive, and attainable objective. First, the decision to end the conflict once Iraqi troops were expelled from Kuwait was a clearly defined objective. The war aim, as agreed at the United Nations when the allied coalition was formed, was not to occupy Iraq, replace the present Iraqi government, or govern Iraq during a transition period to another regime.

President Bush complied with the United Nations resolutions authorizing the collective security action and the limited goals embraced by the whole US-led coalition. To go further might have led to a split in the coalition and would have been on uncertain legal grounds.

Second, despite the US decision to halt Desert Storm operations short of a ground occupation of Iraq, the campaign was, nevertheless, decisive in securing the liberation of Kuwait and in inflicting a decisive defeat and surrender of all Iraqi forces stationed outside of Iraq's borders.

Third, President Bush's objective in the Gulf conflict was quite attainable. Not only was Kuwait liberated, but, after three years, the Iraqi parliament has finally agreed to drop claims to Kuwaiti territory and recognize the borders of Kuwait as legitimate.

President Bush's decision to keep to such clearly defined, decisive, and attainable objectives was determined by the calculation that to go further and invade Iraq would have gone beyond the UN resolutions authorizing the collective action. Such action, it was thought, would endanger the support of coalition partners needed to legitimize the subsequent peace arrangements and whose support the United States would need to guard its interests in future dealings in the Middle East and the Persian Gulf. Further, President Bush and his advisers understood the difficulties of conquering Iraq, locating and capturing Saddam Hussein and his subordinate leaders, subduing the remnants of the Iraqi military throughout a country larger than Germany, and governing a hostile population of almost 20 million while seeking to set up a friendly regime.

The Bush administration was also eager to avoid further bloodshed, having just won the victory in Kuwait at a human cost well below what had been predicted for the ground campaign (150 US dead as opposed to predictions that ranged up to 15,000). President Bush and Secretary of Defense Dick Cheney saw entry into Iraq as a quagmire to be avoided and ended the fighting while the allies were well ahead and had attained their immediate stated goals.

Realpolitik may also have been a factor in the United States's decision to stop when it did. Prior to the 1990 invasion of Kuwait, the United States had been more concerned with containing Iranian power rather than Iraqi power in the region. After all, it was Iran under the ayatollahs who seized American hostages at the US embassy in Tehran in 1979, and who was seen as the chief exporter of anti-American sentiments, and who was seen as the chief exporters of terror worldwide. The fact that Iraq, if totally disarmed by the allied coalition, could not offset the expansionist ambitions of Iran was still another argument for not entering Iraq and totally dismantling its military power in 1991.

Finally, it is likely that President Bush and his political advisors also wished to reap the political fruits of an almost total victory in Kuwait as opposed to entering the political minefield of an invasion, extended military campaign, and occupation of Iraq. By stopping when he did, President Bush received an unprecedented 93 percent approval rating in polls of the American public in the aftermath of the war.

The decisive victory, stopped at its apex, also sent an unchallenged message around the world about US military prowess and American willingness to act decisively against aggression when it felt its vital interests were at stake. This enhanced US reputation in the world could be used to deter other would-be aggressors in places like North Korea, the Persian Gulf, and elsewhere. US credibility had never been higher since the end of World War II, a recovery from the years following the Vietnam War.

Given these arguments in support of President Bush's decision to follow limited war aims in 1991, there is still controversy over whether stopping short of Baghdad was an act of wisdom or short-sightedness. Some believe that the allies should have finished the regime of Saddam Hussein when they had the opportunity to act decisively against him. Time has shown that he has a remarkable ability to survive politically in Iraq, and Iraq has been able to reconstitute much of its conventional military capability even under the terms of the truce. Moreover, Iraq retains the scientific base, foreign supplier contacts, potential wealth from its oil reserves, and ambitions for future great-power status.

Once UN sanctions are lifted on Iraq, many believe that country will be back in the WMD business full-scale. Indeed, resurrection of its biological weapons stockpile should be simple since the allies never found it and therefore did not destroy it. Iraq is given two years of full scale effort before it could be at 1991 levels again in its nuclear weapons research, and less than a decade after that before it could join the nuclear weapons club.

Indeed, not to have deposed Saddam Hussein, when the chance presented itself, may be to have defined the US and UN objective too narrowly at the onset of the Gulf conflict since it is arguable whether the US and allied limited actions achieved a lasting end to the Iraqi threat or merely postponed the confrontation with an Iraq armed with NBC weapons and the missiles to deliver them on target.

The symptoms were treated and their effects mitigated, but the disease persists that could be lethal next time to US interests and allies in the Gulf region. One evidence of Saddam's persistent malevolence was the Iraqi-sponsored attempt to kill former President Bush on his visit to Kuwait in 1993. Leaving such an opponent alive and in power is like allowing a rattlesnake to continue to live in your house after it has attempted to kill you once, because you have temporarily milked it of its venom, even though you know it will inevitably produce more in time.

Permitting Saddam Hussein to remain in power to continue to threaten his neighbors and US interests in the region, by stopping at the Iraq-Kuwait border, is analogous to having allowed Adolph Hitler to remain in control of Germany in 1945 because the Allies decided to stop at Germany's borders once German armies had been expelled from the lands that they occupied from 1939–1945.

Given the track record of Iraq, a state that has been at war with its neighbors since its inception, and of Saddam Hussein, whose regime has constantly used murderous violence against its opponents inside Iraq and aggressive war against its neighbors since he took power, there is a high likelihood that the Gulf War will have to be repeated in the future, perhaps against an even more dangerous enemy.

To conclude, as former Secretary of Defense Caspar Weinberger once advised, "if we do decide to commit forces to combat overseas, we should have clearly defined political and military objectives. And we should know precisely how our forces can accomplish those clearly defined objectives."[25] This is a useful guideline, even if it does not precisely tell you what to do and where to draw the line on your prewar and postwar aspirations.

The Principle of Security in Warfare

Good security means the enemy cannot achieve strategic surprise. Such good security increasingly depends on accurate and timely intelligence information to assess the threat and give timely warning of it in an era when hostile and radical opponents are about to acquire the most destructive of weapons.

It has become increasingly difficult to predict the progress of nonnuclear states as they approach obtaining an operational WMD capability. Most of these regimes find it neither in their political, economic, nor military interests to advertise their progress or capabilities.

The international legal norm established by the NPT carries pledges by the nuclear weapon states that they will not attack nonnuclear signatories of the pact and that they will be subject to sanctions if they violate that pledge. Aspiring proliferators might hide behind their signatures on the NPT to gain legal protection against intervention, particularly if the evidence of their developing WMD is ambiguous.

Declared proliferators may also suffer unilateral cutoffs and sanctions by triggering national legislation on the books in the United States and among other states. These laws enforcing international norms prohibiting proliferation also prescribe various penalties for states that break from the ranks. Witness the Pressler Amendment and the trade penalties inflicted on Pakistan as a result of its nuclear weapons program.

Moreover, as Saddam Hussein learned in June 1981 when his Osirak reactor was destroyed by Israeli warplanes, it does not pay to develop WMD in high-profile, easily targeted facilities. Instead, armed with great wealth from his oil revenues, Saddam from 1981–1991 was able to move very close to a nuclear weapons capability following a clandestine approach. This model is the more likely one for aspirant states to follow, namely:

- Pursuing multiple technical paths to NBC weapons;
- Disguising and hiding WMD facilities, some underground;
- Providing disinformation about WMD activities/locations;
- Joining the NPT as a ruse while clandestinely cheating;
- Using third parties to purchase WMD production technology;
- Purchasing dual-use technologies allegedly for another purpose;
- Producing indigenously as many components of WMD as possible;
- Getting prospective contractors to fill gaps in WMD knowledge through the bid and proposal process, sometimes not letting the contract afterwards;
- Buying as much WMD technology and resources on the open market as possible from contractors all too ready to help in return for substantial profits;
- Hiring foreign NBC/missile expertise where local expertise is lacking; and
- Purchasing WMD technology subcomponents, rather than components, and assembling them inside their country to reduce the audit trail.

Like the proverbial iceberg, just the tips of the North Korean and Iranian nuclear weapons programs are visible, and they probably indicate a much larger clandestine program operating out of sight.

The rate of progress may be accelerated by the possibility of transfers of scientific knowledge, highly enriched uranium (HEU) or plutonium (PL), weapons designs, missiles, and nuclear technology from the newly independent states of the former Soviet Union, which have a surplus of underpaid nuclear scientists and technicians, hundreds of tons of HEU and PL, tens of thousands of nuclear weapons, a need for hard currency, and an expanding criminal element with some access to the widespread nuclear facilities of the former superpower.

According to US military doctrine, the United States should

> never permit the enemy to acquire an unexpected advantage. Security enhances freedom of action by reducing friendly vulnerability to hostile acts, influence or surprise . . . . Thorough knowledge and understanding of enemy strategy, tactics, and doctrine and detailed staff planning can improve security and reduce vulnerability to surprise.[26]

Vulnerability to surprise and attack can be reduced by a combination of offensive and defensive measures. Security can be maintained by a mix that includes:

- Keeping allied escalation dominance to deter enemy escalation to first use of WMD;
- Allied counterforce strikes to destroy or reduce enemy WMD assets;
- Allied active defenses to intercept enemy missile or aircraft attacks;
- Use of passive defenses to protect friendly forces from the effects of nuclear, biological, or chemical weapon attacks;
- Any other measures designed to present less lucrative targets to enemy WMD such as dispersion, mobility, maintaining forces outside of enemy missile or aircraft ranges, and introducing supply and reinforcement means that are less vulnerable to NBC strikes.

No single approach may neutralize the WMD threat, but taken in combination, these measures may greatly reduce the vulnerability of friendly forces to NASTI surprises.

Improved allied capabilities to remotely detect adversary nuclear, biological, chemical and missile assets on the ground or en route to target, would also enhance security and help avoid rude and devastating surprises by the enemy.

The Principle of Economy of Force in Warfare

Another principle of war set out in US military doctrine is to "allocate minimum essential combat power to secondary efforts."[27] In other words, it is recommended that the US commander should concentrate the majority of his military power toward a clearly defined primary threat rather than compromise the effort against secondary priorities. This principle of war is based on the premise that the CINC will not have unlimited resources and must accept some calculated risks in secondary areas of importance in order to achieve superiority in the priority area where the battle or conflict may be decided.

On the grand strategic level, the United States has adopted a strategy of preparing to fight two nearly simultaneous major regional conflicts at the same time. Clearly, utilizing the principle of economy of force, the United States would need to hold in reserve enough force for a second MRC once the first one begins.

The principle of "economy of force" also would serve as a guide to cutting back on secondary US military participation such as in on-going UN peace operations in other regions—so long as US forces are engaged in one or more major regional conflicts, or lack the military power to predominate in both.

The principle of economy of force must be applied with a caveat when an enemy is equipped with WMD. The allied commander must avoid having his main thrust trumped by the employment of enemy mass destruction weapons. Therefore, while resources must be focused on the decisive weak points in adversary forces and plans, they must simultaneously be adequately protected by maintaining intrawar escalation dominance, and their employment prefaced by an air campaign designed to substantially eliminate an enemy WDM capability.

The main ground thrust against enemy forces must be adequately protected by concentrating active and passive defense assets on behalf of the main effort. Dispersion and continued mobility of key force elements, combined with rapid supply and reinforcements from diverse logistics pathways, all done with dispatch, air cover, and secure and clandestine movements of troops, equipment, and supplies, will help preserve the element of tactical surprise and disguise where the main effort will be made.

As US Army and Air Force doctrine states, "economy of force missions may require the forces employed to attack, to defend, to delay, or to conduct deception operations."[28]

The Principle of Surprise in Warfare

US military doctrine teaches that commanders must attempt to "strike the enemy in a time or place, or in a manner, for which he is unprepared."[29] Surprise can affect the outcome of battles, campaigns, or even entire wars. Surprise can be achieved by speed of attack and maneuver, taking unanticipated actions, using deception, varying the tactics used from those previously employed, maintaining operations security, gaining good intelligence and insights into enemy thinking and doctrine, and applying new technologies in ways that reduce enemy warning time, provide capabilities he does not anticipate, or contribute to his confusion.

In the realm of new technologies to achieve surprise, note the importance of stealth F-117 fighter-bombers in striking key targets in the 1991 Gulf War and the use of precision-guided cruise missiles with very small radar cross-sections. One of the architects of the US air campaign in the 1991 Persian Gulf War has written that

> for the first time in the history of warfare, a single entity can produce its own mass and surprise . . . . Surprise has always been one of the most important factors in war—perhaps even the single most important because it could make up for the deficiencies in numbers. Surprise was always difficult to achieve because it conflicted with the concepts of mass and concentration. In order to have enough forces available to hurl enough projectiles to win the probability contest, the commander had to assemble and move large numbers. Of course, assembling and moving large forces in secret was quite difficult, even in the days before aerial reconnaissance, so the odds of surprising the enemy were small indeed. Stealth and precision have solved both sides of the problem; by definition, stealth achieves surprise, and precision means that a single weapon accomplishes what thousands were unlikely to accomplish in the past.[30]

Until the NASTI regimes acquire radars or other sensors capable of detecting and targeting incoming stealth aircraft and cruise missiles, the United States and its allies have a means of achieving tactical surprise in any air strike or any cruise missile launch. The ability to strike "out of the blue" without warning, provided by the B-2, F-117s, future F-22s, and stealthy cruise missiles is limited only by how successfully US and allied intelligence can identify and locate significant enemy targets, and by the availability of stealth aircraft or cruise missiles.

Technological surprise can also decide battles when one side first employs a decisive new military technology which puts the adversary at an unanticipated disadvantage. One of the most dramatic illustrations of this was the decisive role of British radars in helping the Royal Air Force win the Battle of Britain against the German Luffwaffe. Although greatly outnumbered in aircraft, the British were able to pinpoint the directions and numbers of German aircraft as they took off in France and flew across the English Channel toward Britain. Armed with this knowledge, British Spitfires waited high in the clouds in ambush and concentrated superior forces in the air battles they chose to fight. The result was a British victory where bean counters would have predicted defeat. Radar was the biggest difference in the two sides.[31]

The Principle of Simplicity in Warfare

US Army commanders are taught to prepare "clear, uncomplicated plans and clear, concise orders to ensure thorough understanding."[32] Simplicity of operational concepts and goals should reduce misunderstandings of orders, reduce confusion, and enhance the understanding of key audiences whose support is necessary to conduct the war.

This simplicity of operation should be applied to all phases of combat; during the opening phase of operations, in the main campaign, and in the war-termination phase. The war plan should be a continuation of politics by other means, keeping in mind the national ends for which the conflict was begun, constantly relating national ends to ongoing military means, and understanding the unique limits on war termination imposed by the stark fact that the adversary possesses weapons whose destructive magnitude exceeds anything previously faced by other US commanders in previous conflicts.

Of course, simplicity and clarity of goals, plans, and orders alone do not guarantee a correct strategy or successful operation against a heavily armed regional enemy. A CINC could choose a clear, simple plan based on tried-and-true principles, but find that it would not work in a future MRC where the adversary was equipped with radically different capabilities well beyond those possessed by enemies in the past.

Armed with WMD, such adversaries might follow an escalatory strategy that could shatter the cohesiveness of an allied coalition, could scare off potential allies, might inflict a political defeat on the coalition by dissolving allied domestic support for the war, or even cripple an allied expeditionary force by turning NBC and missile assets against allied forces, ports, air bases, logistical tail, or allied capitals in the region. In such a campaign, a NASTI attack might conceivably inflict in a single day allied war deaths in excess of what the United States suffered in Korea, Vietnam, or even in World War II.[33]

The right operational plan will be essential against NASTIs on the field of battle. Clarity and simplicity added to a sound approach contribute to success. Of course, if added to a flawed concept of operations, clarity and simplicity cannot avert defeat.

Additional Principles of War against Enemies with WMD

Military experience and recent technical innovations have spawned some additional principles of warfare to add to the list supplied by General Fuller in World War I. These new operating principles, when combined with the original MOSSCOMES principles of war may supply the decisive edge against radical hostile regimes armed with WMD.

These new principles can be summarized by the acronym SLIP:

S - Simultaneity and Depth of Attack
L - Logistics
I - Information Dominance
P - Precision Targeting

Simultaneity and Depth of Attack

When battling a NASTI, it is best to strike fast and simultaneously at all key enemy assets to stun and paralyze his forces to defeat them in the shortest time possible. Simultaneous strikes throughout the entire battlespace may be enough to rob him of much or all of his WMD capability, and reduce his offensive potential.

As the US Army Training and Doctrine Command states in its concept of operations for the early twenty-first century, "The relationship between fire and maneuver may undergo a transformation as armies with high technology place increasing emphasis on simultaneous strikes throughout the battle space. Maneuver forces may be massed for shorter periods of time."[34]

Army doctrine also notes that "depth and simultaneous attack may be a key characteristic of future American military operations. These operations will redefine the current ideas of deep, close, and rear."[35] Indeed, such parallel war or hyperwar strikes blur the distinction between the strategic, operational, and tactical levels of operations and tend to blend them into one.

Recent effectiveness of simultaneous operations conducted across the full length, breadth, and height of the battle space have led to quick victories in Grenada, Panama, and the Gulf. Desert Storm, for example, showed that "deep battle has advanced beyond the concept of attacking the enemy's follow-on forces in a sequential approach to shape the close battle to one of simultaneous attack to stun, then rapidly defeat the enemy."[36]

Colonel John Warden III, one of the architects of the air campaign that defeated Iraq in the 1991 Gulf War, suggests that near-simultaneous parallel warfare strikes against key enemy leadership, system essentials, infrastructure, population centers, and fielded military forces may impose strategic or operational paralysis on him, leading to his rapid defeat.[37] Warden notes the impact of the fast-paced US parallel air strikes on the 1991 defeat of Iraq:

> In Iraq, a country about the same size as prewar Germany, so many key facilities suffered so much damage so quickly that it was simply not possible to make strategically meaningful repair. Nor was it possible or very useful to concentrate defenses; successful defense of one target merely meant that one out of over a hundred didn't get hit at that particular time. Like the thousand cuts analogy, it just doesn't matter very much if some of the cuts are deflected. It is important to note that Iraq was a very tough country strategically. Iraq had spent an enormous amount of money and energy on giving itself lots of protection and redundancy and its efforts would have paid off well if it had been attacked serially as it had every right to anticipate it would. In other words, the parallel attack against Iraq was against what may well have been the country best prepared in all the world for attack. If it worked there, it will probably work elsewhere.[38]

Thus, the experience of the 1991 Gulf War is that parallel warfare can be decisive since regional adversaries are likely to have a relatively small number of vital strategic targets, estimated by Colonel Warden at "in the neighborhood of a few hundred with the average of perhaps 10 aimpoints per vital target."[39] These enemy assets "tend to be small (in number), very expensive, have few backups, and are hard to repair. If a significant percentage of them are struck in parallel, the damage becomes insuperable."[40]

Of course, there may be countermeasures that an adversary might take to offset the possibilities of simultaneous allied air strikes across the battlespace of a major regional conflict.[41] Efforts might be taken to (1) disguise, diversify and "demassify" the key political-military-economic assets to make them less lucrative targets, (2) hide, harden, or put on mobile launchers, WMD assets to reduce their vulnerability, (3) employ WMD against allied bases from which parallel attacks are being launched, (4) attack allied C^4I and employ various forms of "info war" to confuse, disorganize, and mislead allied commanders and "psychological warfare" to reduce allied morale and influence the publics of the United States and its allies to undermine political support for the war.

Logistics

When drafting the original list of principles of war, General Fuller failed to identify the overwhelming importance of effective logistics to the support of fighting forces as they mobilize, deploy, maneuver, reconstitute, withdraw, and demobilize. Without proper logistics it would be impossible to man, arm, fuel, fix, move, or sustain the soldier, sailor, or airman and their equipment as they enter and fight major regional conflicts. As one US Army general has put it, "Forget logistics and you lose."[42] On more than 230 occasions, US forces have been sent to other countries and regions of the world in the twentieth century alone. Logistics gets them there, sustains them, and gets them home again.

Increasingly, logistics will play an important part in whether US and allied forces get to the battle in time and whether they will predominate when they arrive. This is especially true now that fewer US troops are stationed abroad while still responsible for standing ready to win two near-simultaneous major regional conflicts (MRCs) and participating in a number of military operations other than war (MOOTW) as well.

MRCs and MOOTWs both require a force-projection logistical system that has "the demonstrated ability to rapidly alert, mobilize, deploy and operate anywhere in the world."[43] As a recent analysis of the US Army in the Gulf War notes, logistics units do more than sustain forces in the field. Indeed, "the strength of the logistics engine determines the pace at which an intervening force makes itself secure."[44]

One student of that conflict has observed:

> The Iraqi Army stood by and watched on television as the American Army assembled a sophisticated combat force in front of them with efficiency and dispatch. The act of building the logistics infrastructure during Desert Shield created an atmosphere of domination and a sense of inevitable defeat among the Iraqis long before the shooting war began. In the new style of war, superior logistics becomes the engine that allows American military forces to reach an enemy from all points of the globe and arrive ready to fight. Speed of closure and buildup naturally increases the psychological stature of the deploying force and reduces the risk of destruction to those forces that deploy first. In contrast, dribbling forces into a theater by air or sea raises the risk of defeat in detail.[45]

A successful buildup of US and allied forces and supplies at the inception of a major regional conflict could, in turn, depend upon the early deployment of an effective multilayered air and missile defense and air superiority over the battle zone. As Col Warden has warned, surface forces and logistical support units are fragile at the operational level of war, especially against highly armed challengers.

> Supporting significant numbers of surface forces (air, land, or sea) is a tough administrative problem even in peacetime. Success depends upon efficient distribution of information, fuel, food, and ammunition. By necessity, efficient distribution depends on an inverted pyramid of distribution. Supplies of all operational commodities must be accumulated in one or two locations, then parsed out to two or four locations, and so on until they eventually reach the user. The nodes in the system are exceptionally vulnerable to precision attack.[46]

In short, while the United States and its allies may be able to handle a NASTI regime such as Iraq in 1991, in the future it may be dealing with adversaries that have mastered the building of accurate ballistic missiles, nuclear warheads, chemically armed reentry vehicles, and relatively cheap, hard-to-detect cruise missiles. At that point, MRC forces and their logistics tails had better reduce their vulnerabilities by application of deterrence, preemptive strikes, defenses, deployment outside of enemy range, dispersion of units, constant mobility, or diversity of supply paths in order to avoid defeat.

Information Dominance

The importance of winning the information war should be a guiding principle of wars of the future. A US Army study predicts that "effective information operations will make battlespace transparent to us and opaque to our opponents."[47] Such, at least, is the goal.

One of the air commanders of the Gulf War also emphasizes the importance of information at the strategic and operational levels. He notes that

> In the Gulf War, the coalition deprived Iraq of most of its ability to gather and use information. At the same time, the coalition managed its own information requirements acceptably, even though it was organized in the same way Frederick the Great had organized himself. Clear for the future is the requirement to redesign our organizations so they are built to exploit modern information-handling equipment. This also means flattening

> organizations, eliminating most middle management, pushing decision making to very low levels, and forming worldwide neural networks to capitalize on the ability of units in and out of the direct conflict area.[48]

> The information lesson from the Gulf was negative; the coalition succeeded in breaking Iraq's ability to process information, but the coalition failed to fill the void by providing Iraqis with an alternative source of information. Failure to do so made Saddam's job much easier and greatly reduced the chance of his overthrow. Capturing and exploiting the datasphere may well be the most important effort in many future wars.[49]

Precision Targeting

Another principle of war flowing from technical innovations is the dominance imparted by using precision guided weapons. Suddenly, with great precision, nearly all important fixed targets can be destroyed in a campaign. Instead of having to fire thousands of bombs and missiles at targets, just a few will do the job today with much greater certainty than the imprecise massed attacks of yesterday.

Now "one bomb, one target destroyed" is more the norm instead of "hundreds of bombs, perhaps few or no targets destroyed." This helps in planning a successful campaign and in executing it. MRC logistics are simplified since a finite number of precision weapons can now be used to destroy a set of targets rather than the massive quantities of "dumb" weapons that would otherwise be needed to accomplish the same mission.

The combined advantages of stealth technology and precision guided missiles can be seen by comparing a conventional bombing attack in the 1991 Gulf War, against the same target, the Baghdad Nuclear Research Center, with a stealthy precision attack two days later. The conventional air attack failed to destroy the target even though it used 32 bomb-dropping aircraft, 16 fighter escorts, 12 aircraft for suppression of Iraqi air defenses, and 15 tanker aircraft. Two days later, this target was successfully destroyed using just eight F-117 stealth fighter-bombers supported by just two tankers.[50]

Conclusions

There are many other principles of war that might be formulated to apply to different kinds of engagements. For example, war against a NASTI is far different from participating in military operations other than war such as UN peace operations. Further, low-intensity counterguerrilla warfare is prosecuted differently than more conventional battle, as fought in the 1991 Gulf War, and both might be fought differently in future wars.

One scholar has listed over one hundred principles of war that have been advocated by military thinkers since the time of Sun Tzu.[51] Indeed, in 1984 US Air Force doctrine recommended four guidelines (timing, tempo, logistics, and cohesion) in addition to Fuller's original list of nine principles of war. Some in the recent past have argued for the inclusion of the concept of deterrence as a separate principle of conflict management.[52]

A review of the principles of war that pertain to a future conflict with an enemy equipped with advanced conventional arms and mass destruction weapons can provide a better understanding of how to operate on the future battlefield. However, such a set of principles are not infallible guides to action. They cannot substitute for judgment, improvisation, insights into the enemy, or initiative. Nor can they be applied by rote or as part of a checklist.

Understanding of these principles can add to the commander's understanding of how to conduct warfare, and a review of them can remind him of fundamentals to observe, but such application of these principles by themselves is not sufficient for victory. For one thing they are somewhat abstract and require judgment in application to specific cases.

In the end, the commander and his subordinates still must bring a depth of experience, concrete mastery of details, and an understanding of military affairs that reaches well beyond such general principles. Nevertheless, these principles of war can be useful ways to think about how to solve the problem facing a commander whose force is opposed by a NASTI.

Notes

1. Rogue states whose military forces are equipped with WMD and means of delivering them on targets may interfere with highly sophisticated new strategies and technologies developed by the United States and its allies. WMD may level the playing field. For example, a high-altitude nuclear EMP burst may destroy allied communications, interfere with space-based reconnaissance, impede the digitalization of the battlefield, blind allied precision strike forces to new targets, and serve as a form of information warfare in its crudest form. The RMA brought about by the introduction of NBC and missile systems into a theater of war may predominate over the effects of other strategies and technologies. For a different view see the chapter on "The Revolution in Military Affairs" by Jeffrey McKittrick, James Blackwell, Fred Littlepage, George Kraus, Richard Blanchfield, and Dale Hill in this volume.

2. J. F. C. Fuller, "The Principles of War, with Reference to the Campaigns of 1914–1915," *Journal of the Royal United Service Institution*, vol. 61, February 1916. Fuller cited seven principles of war: Objective, Offensive, Mass, Economy of Force, Surprise, Security, Cooperation. Later, the US military dropped Cooperation as a principle of war and substituted Simplicity and Command Unity.

3. See Air Force Manual (AFM) 1-1, *Basic Aerospace Doctrine of the United States Air Force*, March 1992, vol. 1, 1 and 2, Essay B, "Principles of War," 9–15. See also, Appendix A, "Principles of War," Army Field Manual (FM) 100-5, Operations, 1990 edition, 173–77.

4. Ibid., 174. The principle of mass can also be applied at the grand strategic level. "In the strategic context, this principle suggests that the nation should commit, or be prepared to commit, a preponderance of national power to those regions or areas of the world where the threat to national security interests is greatest." See also AFM 1-1, 1.

5. The advantages of achieving local superiority, even if outnumbered overall, is not a new idea. The same point was made several thousand years ago by Sun Tzu. See Sun Tzu, *The Art of War*, trans. Samuel B. Griffith (London: Oxford University Press, 1963), 98.

6. The Manhattan Project provided a fission weapon in 1945 that was over a thousand times more powerful per unit weight than a TNT warhead of equivalent weight. The H-bomb fusion weapons that followed carried an explosive yield a thousand times more powerful than the earlier A-bombs. This millionfold increase in explosive capability between 1945 and 1950 was augmented by the first-time capability to deliver such weapons across intercontinental distances by aircraft and missiles. Massed local forces were now targetable by WMD delivered across intercontinental ranges.

7. Gen Gordon R. Sullivan and Col James M. Dubik, USA, "Land Warfare in the 21st Century," *Military Review*, September 1993, 22. Their chart on "The Expanded Battlefield" traces the density of troop deployment, width of the battlefront, and depth of the battle space in wars from antiquity to the 1991 Gulf War. The earlier work done on wars of antiquity, Napoleonic wars, the American civil war, World War I, World War II, and the

October War was found in Col T. N. Dupuy, *The Evolution of Weapons and Warfare* (Fairfax, Virginia: Hero Books, 1980).

8. Michael Mazaar, "The Revolution in Military Affairs: A Framework for Defense Planning," Strategic Studies Institute, US Army War College, Carlisle Barracks, Pennsylvania, June 10, 1994, 20.

9. Lt Col Edward Mann, USAF, "One Target, One Bomb: Is the Principle of Mass Dead?" *Military Review*, September 1993, 33–41.

10. Ibid., 16.

11. Ibid., 18.

12. Ibid., 21. Indeed, Mazaar notes, more broadly, that "developments in warfare are reducing the role of major platforms—heavy ground vehicles, large capital ships, and advanced aircraft." See also Adm David E. Jeremiah, "What's Ahead for the Armed Forces," *Joint Force Quarterly*, no. 1 (Summer 1993); 32.

13. Mann, 37.

14. Army, FM 100-5, Appendix A, 1990, 173.

15. Army Chemical School, "Summary Evaluation: Report for Combined Arms in a Nuclear/Chemical Environment (CANE) Force Development Test and Experimentation, Phase 1, March 1986. This source was cited in an article by Maj Gen Robert D. Orton, USA, and Maj Robert C. Neumann, USA, "The Impact of Weapons of Mass Destruction on Battlefield Operations," *Military Review*, December 1993, 66.

16. Ibid., 68.

17. Ibid., 68–71.

18. US Army, Training and Doctrine Command (TRADOC), "Force XXI Operations: A Concept for the Evolution of Full-Dimension Operations for the Strategic Army of the Early Twenty-First Century," *TRADOC Pamphlet 525-5*, August 1994, 3–21.

19. AFM 1-1, vol. 1.

20. Ibid., 5.

21. It would be inadvisable in almost all contingencies to use US nuclear weapons in counterforce strikes against enemy weapons of mass destruction. For a number of reasons, the preferred instrument for disarming the adversary very likely should be advanced conventional weapons. The worldwide reaction to the United States using nuclear weapons on a regional enemy, particularly if used first, would be negative in the extreme and could unhinge all other US diplomatic and multilateral efforts to counter the spread of nuclear, biological or chemical weapons in other regions of the world. For one thing, nuclear strikes against a state that had signed the NPT are illegal under the treaty, so such an action would be a flagrant violation of international law. Indeed, any state that violates this code would be taking a course diametrically opposed to UN Security Council pledges to punish such violators. Indeed this has historically been a US position at the United Nations to punish nuclear first use against NPT members. Further, US nuclear first use in a regional war, even against a NASTI, would undoubtedly arouse world opinion against US policy, making the United States a pariah state in many more quarters. It would not be unexpected to see Americans and US property assaulted all around the globe in retaliation. Moreover, US nuclear first use, even against a NASTI, would shatter the nonnuclear international taboo that the United States has

attempted to foster with treaties and diplomacy for decades. Finally, such a policy of nuclear first use could also cause a collapse of US domestic support for the regional war effort that would probably rival or exceed the antiwar activities inside the United States during the Vietnam War period.

22. "Force XXI Operations," 2–8.

23. During the cold war, one fear of US strategists was the nuclear decapitation strike from the Soviet Union, perhaps by an off-shore sea-launched ballistic or cruise missile. See Barry R. Schneider, "Invitation to a Nuclear Beheading," *Across The Board*, 20, no. 7 (July/August 1983): 9–16.

24. If the first engagement is decisive enough, the conflict may be over almost before it has begun. This was true, for example, of the United States's intervention against the Noriega regime in Panama.

25. Caspar W. Weinberger, "The Uses of Military Power," text of remarks by the secretary of defense to the National Press Club, November 28, 1984. This is included in the appendix to Weinberger's book, *Fighting for Peace: Seven Critical Years in the Pentagon* (New York: Warner Books, 1990) 441.

26. FM 100-5, "Appendix A: Principles of War," 176. For a similar commentary, see the June 1993 edition of FM 100-5, 2-4 to 2-6.

27. Ibid., 174–75.

28. Ibid., 175.

29. Ibid., 176.

30. John A. Warden III, "Air Power for the Twenty-First Century," in Karl P. Magyar, Editor in Chief, *Challenge and Response: Anticipating US Military Security Concerns* (Maxwell AFB, Alabama: Air University Press, August 1994) 328–29.

31. See Chester Wilmot, "David and Goliath," chapter 2, *The Struggle for Europe* (New York: Harper Colophon Books, 1952,) 33–55. Not realizing the potency and importance of British radar stations, the German high command mistakenly abandoned their early bombardment of them in the Battle of Britain because they believed that the British would be able to repair them and put them back into operation very quickly. Had the Germans persisted, they may have won the air battle over England.

32. FM 100-5, 177.

33. In Korea, the United States was reported to have lost 35,000 troops killed in combat; in Vietnam the number was 53,000 ; and in World War II, 330,000.

34. "Force XXI Operations," 2–9.

35. Ibid., 3–11.

36. Ibid., 2–9.

37. Warden, 311–32.

38. Ibid., 325.

39. Ibid., 327.

40. Ibid.

41. Col Richard Szafranski, USAF, "Parallel War and Hyperwar: Is Every Want a Weakness?" See elsewhere in this volume.

42. Gen Frederick M. Franks, Jr., as quoted in Col Michael S. Williams and Lt Col Herman T. Palmer, USA, "Force-Projection Logistics," *Military Review*, June 1994, 29.

43. FM 100-5, June 1993, 3–6.

44. Gen Robert H. Scales, USA, *Certain Victory: The U.S. Army in the Gulf War* (Fort Leavenworth, Kansas: US Army Command and General Staff College Press, 1994), 378.
45. Ibid., 376.
46. Warden, 328.
47. "Force XXI Operations," 3–21.
48. Warden, 329–30.
49. Ibid.
50. Mann, 38.
51. John I. Alger, *The Quest for Victory: The History of the Principles of War* (Westport, Conn.: Greenwood Press, 1982). Alger drew on the military writings of thinkers such as Sun Tzu, Machiavelli, Jomini, Mahan, Rocquancourt, Steele, MacDougall, Liddell Hart, Mao Tse-tung, Montgomery, as well as the British army, French army, German army, and US Army and Air Force in compiling his list of principles of war. These guidelines included, for example, diverse maxims on the need for cooperation, shock, favorable ground cover, vitality, fire superiority, flexibility, an indirect approach, simultaneity, reconnaissance, local superiority, air superiority, a will to win, readiness, pursuit, God's blessing, and the moral high ground.
52. For a summary discussion see AFM 1-1, vol. 2, Essay B: "Principles of War," 14. The original sources are Col Robert H. Reed et al., "On Deterrence: A Broadened Perspective," *Air University Review*, May–June 1975, 2–17, and John M. Collins, "Principles of Deterrence," *Air University Review*, November–December 1979, 17–26.

Overview: New Era Warfare? A Revolution in Military Affairs?

Are we currently living through a "revolution in military affairs?" Several of our authors in this book argue that this is the case. What is a revolution in military affairs (RMA)? For our purposes, an RMA is defined as a *fundamental* change, or discontinuity, in the way military strategy and operations have been planned and conducted.

Sometimes an RMA is driven by *technological innovation*, such as the introduction of nuclear weapons at the end of World War II. Sometimes *operational innovations* change warfare—for example, the German blitzkrieg. *Societal changes* such as Napoleon's *Levée en Masse* also can contribute to RMAs. At other times, an RMA may be created by a *combination* of developments—indeed, the reinforcement and integration of military operational, technical, and even socio-economic developments.

The essay by Jeffrey McKitrick, James Blackwell, Fred Littlepage, George Kraus, Richard Blanchfield and Dale Hill, analysts at the Science Applications International Corporation (SAIC), argues that we have been in the midst of an integrated RMA since the start of the Gulf War and that it is accelerating. The SAIC analysts contend that the current RMA is characterized by an integration of operational, organization, and technical capabilities across all operating mediums—air, land, and sea. Furthermore, they argue that "new warfare areas," such as long-range precision strikes, information warfare, dominating maneuver, and space warfare, are also emerging.

Regarding deep precision strike, there have been dramatic increases in the ability to strike strategic targets. "In 1943," writes the SAIC team, "the US 8th Air Force prosecuted only 50 strategic targets during the course of the entire year. In the first 24 hours of Desert Storm, the combined [coalition] air forces prosecuted 150 strategic targets—a thousand-fold increase over 1943 capabilities." In 25 more years, it may be possible to strike as many as 500 to 1000 targets in the *first few minutes* of a campaign, inflicting strategic paralysis on the enemy.

Information warfare constitutes another revolutionary aspect of the new era warfare—quickly disrupting or destroying an adversary's command and control system, intelligence, information propaganda abilities, and general situational awareness.

"Dominating maneuver" refers to the positioning and employment of forces anywhere in a theater so that, in combination with precision strike and information warfare, an adversary's center of gravity can be destroyed.

Space warfare involves dominating the "high ground" of space to deny its advantages to the adversary and to use it to implement one's own command, control, communications, navigation, reconnaissance, air defense, missile defense, warning, and weather forecasting. Space assets can become a key to the future digitalization of the battlefield where some of the fog and friction of war is removed for the side dominating space. Space warfare includes aspects of the other three emerging warfare areas, but has the potential to become a distinct warfare area of its own. The SAIC analysts conclude that truly revolutionary effects in warfare will occur when two or more of the new warfare areas combine. The effect may be to so dominate an adversary before conflict starts as to make the conflict unnecessary, something Sun Tzu advocated nearly 2,500 years ago.

Gen Charles Horner also believes we have entered a period where new technologies can change the manner in which we fight wars and can downgrade the value of possession of nuclear arms. This officer, who led the US and allied air operations in the 1991 Persian Gulf War, writes that

> Desert Storm represents a revolution in warfare. Specifically, we will need to conduct operations in ways that inflict the minimum number of casualties on both sides. Additionally, we must prepare for wars in which ballistic missiles are used against our own troops and as terror weapons.... In the end, we should aim to reduce our nuclear arsenals to zero as we substitute missile defenses for nuclear weapons.

Horner's essay concentrates on the need for effective theater missile defenses (TMD) and strong strategic ballistic missile defenses (BMD). The author notes how ballistic missiles and their technology are proliferating across the third world, and have fallen into the hands of rogue states.

Horner argues that multilayered ballistic missile defenses will work better than single terminal-phase systems such as

Patriot, which was the only TMD system available in the Gulf War. Multilayered defenses are designed to intercept ballistic missile attacks all along their flight trajectories. For example, such defenses will interdict the attack at several key points:

- Boost phase and post-boost phase intercepts—the best time to hit a missile because "it's signature is high, it's at its slowest speed, it's vulnerable, and if you get it then, it falls on the enemy." Furthermore, a single intercept of a MIRVed system kills all the weapons with a single shot, before the reentry vehicles carrying the warheads can separate and are en route to individual targets.

- Midcourse intercept—also desirable because one can shoot at the missile in a space environment that interferes less with high-energy defensive weapons, and can target the reentry vehicles that escaped the boost-phase defenses.

- Terminal-phase intercept—more dangerous, because the debris can fall on the defender. This last line of defense has less time to react to the incoming reentry vehicles and is presently restricted to kinetic kill or explosive techniques for intercepts since the earth's atmosphere interferes with laser and particle beam projection. However, if the first two layers of defense do their job well, the attack will be greatly diluted and will present fewer targets for ground-based BMD interceptors.

A two-layered theater missile defense system is currently planned by the United States for future major regional conflicts (MRCs), perhaps the Theater High-Altitude Area Defense (THAAD) system in combination with an improved Patriot terminal defensive system, to help protect US and allied forces in future MRCs.

General Horner advocates a strong, fielded ballistic missile defense system for the United States, our allies, *and* Russia, which is surrounded by would-be proliferators. However contends that the United States could begin more extensive cooperation with the Russians by sharing information on warning systems, and, as we build trust, might cooperate in constructing shared missile defenses. This controversial view of one of the nation's foremost military leaders has attracted much debate.

Chapter 2

New-Era Warfare

Gen Charles A. Horner, USAF (Ret.)

In the twenty-first century we are going to confront increasingly the threat of ballistic missiles and the need for ballistic-missile defenses.

The threats posed by ballistic missiles are obvious. Many nations now have them, and not all those nations are high-tech. There is also no doubt about the fact that the capabilities of ballistic missiles are increasing. Former CIA Director Woolsey, for example, noted that North Korea has developed three new ballistic missiles.

But our real problem is not so much the threat itself, but a lack of understanding about what is the threat. The military missed the whole point in Desert Storm. We used to look upon ballistic missiles in terms of warhead size and accuracy—circular error probability and the like—and calculate the warhead's impact on the enemy, but we missed the real impact of the ballistic missile.

Part of the reason for this misunderstanding is simply our heritage. All of us here today have lived the majority of our adult lives during the cold war, which caused us to look at things in strategic terms; that is, in the context of US/Soviet relations. We still argue, for example, about the futility of having ballistic missile defenses because it's bad to defend against mutual assured destruction. But many of our views on ballistic missiles, couched in the obsolete terms of the cold war bipolar world, are not appreciated by much of the world today. That is particularly true of Saudi Arabia, Bahrain, and Israel.

Edited remarks originally delivered by Gen Charles A. Horner on 19 July 1994 at the Washington Roundtable on Science and Public Policy, sponsored by The Marshall Institute. His speech is reprinted here with the permission of the George C. Marshall Institute, Washington, D.C.

There is also a history we can refer to regarding the use of ballistic missiles, and I was puzzled during Desert Storm that we didn't seem to be aware of that history. Ballistic missiles were used against Rotterdam and London in World War II. After the Gulf War, I was watching a story about our stealth fighter, the F-117. The person largely responsible for the stealth technology on the F-117 is an Englishman who had grown up in London during WWII. He said, "I recall the terrorizing impact of the V-2 attacks. They were far more fearsome than bomber attacks." So the lesson was there that ballistic missiles can have a large psychological impact as well as a military impact on a war, but we had failed to learn that lesson.

The ballistic missile had a profound impact on the coalition nations in Desert Storm, on our forces, and on our understanding of the ballistic missile's own utility. The ballistic missile was the only advantage that Saddam Hussein had in that war. That's the lesson that Saddam taught us, that ballistic missiles may have little military value but do have great terror potential.

Ballistic missiles and ballistic-missile defense carry heavy political implications. For example, if we have no money to spend on conventional defense, we tell our people that it doesn't matter, because we have ballistic missiles, and that since defenses against missiles are destabilizing, it is not even in our best interests to spend money on a defense against them.

But in reality ballistic missiles *are* with us, and the capability for launching them is growing. Technology transfers, a technology revolution occurring every 18 months to two years, as well as legitimate space launches, are things we need to be concerned about in this regard. In today's world there is no reason why any nation can't put a communications satellite into space. That same capability for launching satellites can be transformed, with proper guidance, into a ballistic missile of intercontinental range. Other technologies enhance this delivery capability, such as our Global Positioning System (GPS), which gives nations the ability to target with some degree of accuracy without having to go through the same costly and intense development that we had to go through to get that system operational. There are also many nations that are willing to use ballistic missiles. Iraq, Yemen, and Afghanistan are all examples. We've seen testing on the part of Vietnam, Korea, and Syria, and they are seen as necessary for defense by countries such as Saudi Arabia and Israel. Certainly, there are opportunities we can foresee for countries such as Libya and North Korea to use ballistic missiles; and they are the cornerstone of defense for the United States and Russia, and also for countries such as India and Pakistan.

The world has changed, and the capability exists for many countries to use ballistic missiles. If you put these facts together with other big shifts, such as the decline in nation-state status, and new stresses such as international migration and environmental degradation, we must understand that the world is changing. Certainly there are new strategies being pursued. There is no doubt, for example, that many states are pursuing a strategy of acquiring weapons of mass destruction and the means to deliver them. This is the war that is replacing the cold war.

Lessons of Desert Storm

So let me turn to Desert Storm because it illustrated this revolution in warfare. First, look at this example. At a Camp David meeting in September 1990, before the Gulf War began, President Bush asked: "How do we avoid casualties?" At first, I thought he was referring to US casualties, and then I thought he meant casualties of all the possible allies, but as he continued to speak I realized he was talking about the Iraqis. He was asking how we could conduct this military operation with minimum casualties on *both* sides. That is part of new-era warfare.

There's no doubt that casualties in modern warfare—at least from our point of view—are quite unacceptable. If the casualty count is too high, you can win the war on the battlefield yet lose it at home because you inflicted large numbers of casualties on the enemy or if they inflicted large numbers of casualties on you. Both are unpopular with the US public.

How did we carry out ballistic missile defense (BMD) in Desert Storm? The answer is, we were ill-prepared, though we did prepare to some extent. In 1988 in an exercise under Gen George Crist (commander-in-chief of Central Command), Army Brig Gen Jim Ellis was given free play as a Red Force commander. He was supposed to challenge us. The scenario was that the Russians attacked through Iran and we had to defend. What Jim Ellis did was to keep firing ballistic missiles into my area. Every time I'd mass my force, he would send a couple of ballistic missiles at us. It's a war game, but I'll tell you, pretty soon you get a little tired because you want to be a hero in front of your boss, and this brigadier general was not cooperating at all. But he did us a great service by making us aware of the military utility of these weapons.

In another exercise in July, 1990, the scenario was that Country Orange was going to attack Kuwait and Saudi Arabia from the north, and we knew Country Orange probably had ballistic missiles. I said, "We're not going to be unprepared again." So I went down to see Lt Gen John Yeosock, commander, Third US Army, who was the ground force commander, and I told him that I wanted to use the Patriots to defend against ballistic missiles.

At that time, Patriot was believed to have the capability to intercept ballistic missiles, so I took the Patriot air defense circle and put it on my maps. When we plotted those circles, they just about covered the map. Of course, we learned in Saudi Arabia that the Patriot ballistic missile defense circle looks more like the head of a pin.

We have all heard a lot of very smart people talk about whether the Patriot missiles were or were not a success in Desert Storm. I can only tell you that I had to make a decision about whether or not to move the E-3 AWACS (Airborne Warning and Control System) aircraft out of Scud range. They were positioned at Riyadh Air Base and Saddam was shooting Scuds at them. I had to decide whether to leave them there or move them to Thumrait, which would have put another hour or hour-and-a-half between them and their operational area. I decided to leave them at Riyadh and, because of the Patriot missiles, we didn't lose a single AWACS. That was either a brilliant decision or a very lucky one, I don't know. But I do know that from where I stood, the Patriots worked.

After the initial Scud launchings, we saw the value of the Patriots and of defense against ballistic missiles in general. There can be no doubt about the value of these antiweapon interceptors. Consider that the greatest number of casualties came from one Scud attack on the Dhahran barracks.

This could prove to be a lesson not lost on other nations in terms of the terror caused by ballistic missiles and the political leverage they provide. Suppose Iran, for example, had nuclear weapons and ballistic missiles, and suppose they decided that they wanted to raise the price of oil, and they tell their neighbors that they are going to deny them and every other nation access to the Red Sea. They say to the Saudis, "You know we have these weapons, but we're not interested in occupying your country, nor are we interested in crossing into it; we just want to get the price of oil back to where we can make some money." What would be the reaction of the Saudis?

And finally, it's reasonable to believe that the development and proliferation of ballistic missiles in the hands of potential adversaries has reached a stage where they can affect such things as the building and maintaining of coalitions. For example, what if we were involved in a situation like Desert Storm, where an ally like Italy, for example, was threatened with such a missile attack? Would other neighboring countries, also within range, dare to ally themselves with Italy or the United States? How difficult would it be to hold together a coalition as diverse as the one we had in Desert Storm if the member nations were threatened with direct missile attack?

Status of Ballistic Missile Defense Programs

Where are we today with respect to ballistic missile defense? We have the defense guidance on theater, national, and technology demonstrations, and we have the ABM Treaty, which is the most significant element at the policy level. The ABM Treaty made sense in a cold war context, but you wonder if it has merit in the new world. So far as our current capability is concerned, warning of long-range missiles is provided by space-based infrared sensors and radars. But we should keep in mind that our radars are configured for the cold war. Although we can change that fact, as of now our infrared systems for theater defense have limited detection capability. They have difficulty with land masses far away from the equator, with weather, and with the smaller infrared signature of missiles like the Scud. The biggest problem with our current radars and space-based sensors, however, is that they do not allow us to target and attack mobile launchers. That was one lesson I brought back from Desert Storm, and it is the one I keep hitting over and over and over.

We have made some progress since then. We have modified the software associated with the Defense Support Program (DSP) signal. We can take two or more satellites and interweave their coverage to improve accuracy and the likelihood of detection. But they still do not offer the degree of certainty that I think is needed and wanted. Other countries have offered us the use of their some of their warning assets. For example, the head of the Ukrainian Air Forces asked me, "How would you like to have some of our ballistic-missile radar gear, left over from the cold war?" But we looked at its location and decided we were not interested because it really had no potential military benefit. Of course, the Ukrainians are trying to get close to the United States to preclude being overpowered by the Russians.

Among those most interested in our ballistic missile warning capabilities, and how they can be shared, are our allies. The long-term program that I think would be most effective in this respect is called ALARM. It gives us the ability to see lower signature missiles and it gives us the ability to see missiles that are coasting, for example. Finally, ALARM gives us responsiveness, in that it will be on a smaller launch vehicle and maybe on other satellites.

We need to be concerned about keeping our warning capabilities effective. For example, a DSP satellite went bad last year, so I said launch the spares. I thought it might take 90 days to do this. It turns out that it takes considerably longer. In some cases, it can take up to two years to replace a warning satellite.

The United States can count on our offensive missiles, but we have to think in depth. Too often the nature of our acquisition system forces us into being too programming oriented. We must think not in terms of the single solution but in a variety of solutions. For instance, we wanted to attack Iraq's missiles where they were built, before they were on the launch pads. That would have been the most productive approach. We have the capability to do that, but we did not do a very good job at finding and destroying Iraq's missiles in Desert Storm because we really had not thought sufficiently about the problems beforehand.

After the war, I asked several questions about the command and control of Iraq's systems, and we found out that we could have done some things to improve our missile defense efficiency. For example, launch orders in Iraq tended to come from a very high level, so there were vulnerabilities that could have been exploited.

In general, we need to consider the three areas of missile defense: boost-phase, midcourse and terminal. We have other capabilities, such as JOINT STARS, GPS, and multitarget detection, but I think most of us want to talk about what happens after missiles are launched.

Boost-Phase Intercept

During the Gulf War, we had the potential for an intercept of a Scud just after launch. An F-16 was over Iraq suppressing Scuds, and the pilot saw a missile coming off the pad. He thought it was a SAM being shot at him, so he made a move away from it. He then realized that there was no radar warning and that the missile was very bright and large for a SAM. When he realized it was a Scud, not a SAM, he attempted to shoot it down with an AIM-9 heat-seeking missile. But the high rate of acceleration of the Scud was too much for the F-16.

Boost-phase is a good time to intercept a missile because its signature is high, it's at its slowest speed, it's vulnerable, and if you get it then, it falls on the enemy—a beautiful thing. The problem is having the time to detect and launch on it. The timeline on a boost-phase intercept is very, very tight.

Still, there are many options for boost-phase defense. You can get airplanes to cap an attack, as we did during the Gulf War. You can have protection from large airplanes if you want to use an airborne laser, for example. There are also options for putting unmanned airplanes up for long periods of time.

We need to work on the boost-phase intercept issue, but right now it has been hampered. We have three boost-phase programs that I know of, all on airplanes: HARM, AMRAAM, and the airborne laser. But I think that the current budget crisis will probably keep these programs from being acquired and deployed. That we have been taking funds from these programs tells us more about our own lack of perception of the seriousness of the threat than it does about the threat itself.

Midcourse Intercept

The advantage of a midcourse intercept is that during this period the missile has a good signature because it's out there in a sterile, space environment and you have multiple opportunities to shoot at it. Also, if you go to space-based systems for intercept in mid-course, you get wide-area protection; you can cover larger portions of the landmass depending on how many of these interceptor platforms and sensor satellites you want to put in orbit. Of course, the big down-draw on anything in space is the ABM Treaty, and the widely held attitude that we should not have weapons in space.

My answer to that is that the weapon in space is the warhead on a missile. The interceptor is an antiweapon. Therefore I don't have the same philosophical problems. Parenthetically, I'm not in favor of weapons in space.

There are still other options for midcourse defense against ballistic missiles. The Navy Upper Tier is one, because it gives you a great deal of flexibility. You can park these ships anywhere. If we deployed this system you could have a national missile defense by stationing them as far away as Hudson Bay. But again, midcourse intercept requires midcourse guidance, and that involves sensors in space which track cold bodies and relay that data to the intercept missile, and that is prohibited by the ABM Treaty.

Terminal-phase Intercept

Finally, we have terminal defense, where most of our work is now done. The most logical elements would be development of an improved Patriot; Theater High Altitude Area Defense (THAAD), a ground-based system by the Army; and/or Navy Lower Tier.

However, I worry about this. Will these ground-based systems address the future threat? Too often, when people approve building a theater missile defense, they are talking about building a missile defense against a benign nonthreat, a Scud-like threat. Remember, the Scud wasn't even good enough for Saddam. Even if we deploy one of these systems—such as THAAD—we still need a cold-body tracker in space. If I'm sitting here with my THAAD missile and somebody launches a ballistic missile at me, I still need very definite data as to where that missile is so that I can launch my THAAD before knowing where it's aimed. So warning is not enough—we need to have accurate information about where the missile is. A cold-body tracker doubles the range of your ground-based or ship-based systems. But again, to deploy that tracker requires ABM Treaty changes.

Need for a Robust TMD

The technology for building effective ballistic missile defenses is available today. The dollars have been spent, and countries are clamoring for ballistic missile defenses. I believe that we must be concerned about what kind of ballistic-missile defenses we're allowed to build. Those concerns have to be voiced, or else we will wind up with a noneffective system against fair-to-reasonable threats. If space systems can give us the capability we need to defend our men and women when they are overseas, let alone defend the United States, then we need to seek that. If that means we need to work with the Russians to change the ABM Treaty, that's what we need to do.

Let's talk about the ABM Treaty. I believe the ABM Treaty is a cold war hang-over. is the cold war over? If you look at the defense budget, the cold war is over. What's more, I've had visits from the head of the KGB, and Col General Ivanov, head of Russia's rocket forces. We're talking with them about shared exercises; I've been invited to Moscow. Given all this, I think it's reasonable to assume that the cold war *is* over—and, therefore, I think the ABM Treaty has outlived its usefulness.

That does not mean that we and the Russians are not going to have tensions, conflict, disagreements, and competition. But, I think that things have fundamentally changed between our two countries. Yet we still have ballistic missiles pointed at each other. To what end? We are not going to fight each other. So, it's now a question of how we walk away from the cold war, not do we walk away from the cold war.

In walking away from the cold war, one area we must investigate is shared ballistic missile defenses with the Russians. That requires building trust with the Russians. Nor is it an easy thing to do since they are still coming from a bipolar world view, and I think that they are genuinely concerned about our ability to outstrip them in technology. They want to be in the driver's seat in determining how we march forward on ballistic missile defenses, and the ABM Treaty gives them the leverage to do that. I ran into this problem with Ivanov. We were at the national test facility, which is an impressive operation, and we were briefed by people from our Ballistic Missile Defense Organization. You could see that General Ivanov, who is a very thoughtful, intelligent, and tough guy, was threatened. He looked as if he were thinking, "My God, I didn't know they were this good."

I knew we were going to reach this point, so I had spoken with him all morning about how could we cooperate with him. "What we need to do," I said, "is work together on this. We need to share these technologies. You have things that we can't match, like heavy-lift access to space. There's a marriage right there—our technology in ballistic missiles and your access to space to put things in orbit. There's no reason our two countries can't work together, other than the fact that there are those in both of our capitals who are still very close to the cold war." And he said, "What is your problem in achieving this technology?" I said, "Well, it's the budget. We have to fight for these programs before our Congress." When he went to get on the airplane to leave, he turned and said, "Good luck with your Congress."

In both Washington and Moscow there are people who are still operating in a cold war context. I think the Russians are terrified that we are going to get ahead of them in ballistic missile defenses. So, it is very important that we do things, such as share ballistic-missile warning as an entree, and then become involved in shared ballistic-missile defenses. At that point, the ABM Treaty is moot.

How you build trust with the Russians is a matter of a lot of effort, discussion, and dialogue. But I think to approach it from the cold war standpoint is the wrong tack. I think we must approach it looking at things through their eyes as they look to the south, not over the pole. I have no doubts that the Russians are quite willing to do that, given time.

The biggest problem we have is our own lack of understanding of the threat. I'm amazed by that. Many people come to Cheyenne Mountain where we show them the ballistic missile warning systems. Then we ask them, "What do you think of our ballistic missile defenses?" I would estimate that 60 percent of the people say, "They have got to be the finest in the world, and we can't thank you enough." But the truth is we have none.

That is a scary thing—they don't know. The danger here is that if we are threatened by ballistic missiles someday, the American people are going to feel betrayed. So, I think that, while we have no threat for now, we must communicate to the American people that they do not in fact have ballistic missile defenses, and that there is a potential for them at some point in the future to be attacked.

Shared Defense Against Ballistic Missiles

What do we need to do? First, recognize that the world has changed; second, recognize that the cold war is over; and, finally, recognize that we and the Russians have much to gain from getting rid of our nuclear arsenals. Think about this: the nation most threatened by the proliferation of missiles and weapons of mass destruction is Russia. They are surrounded by North Korea, China, Pakistan, India, Syria, Iraq, Iran, Afghanistan, and Ukraine, and all these states have nuclear weapons or nuclear-weapons programs, and all have ballistic missiles or ballistic-missile programs. Russia needs ballistic missile defenses more than does the United States. Why do we not start the process by sharing warning, and as we build trust, then sharing defenses?

What's the cost of sharing warning, for example? It means we give up the opportunity to attack them on a surprise basis. We have people who are fighting shared warning, but I think their arguments are based on cold war fears, not on the world we find today. I think that we have to recognize this world is a dangerous place in many areas, and that proliferation of weapons of mass destruction and the means to deliver them are the true threat that we need to be addressing. Diplomacy is best, and we have nonproliferation regimes, which we need to pursue. But there are determined adversaries out there who are not deterrable by counterforce because, if they are smart at all, they know that we are not going to use nuclear weapons.

What are nuclear weapons good against? They are good against cities. Are we going to bomb the capital of some country in order to deter them from using nuclear weapons? They know we are *not* going to do that. Now, we do have sufficient conventional military strength, and particularly with a coalition as in the Gulf War, we can be effective on the battlefield. But if there is a question about our national will, then it's in some nation's interest to build these weapons of mass destruction and threaten people.

We must be credible on the basis of conventional defense. In Desert Storm, Saddam Hussein had more chemical weapons than I could bomb. We had to make a decision to go after the production areas and those storage areas that were in the immediate area of the battlefield. I could not have even begun to take out all of his chemical storage—there are just not enough sorties in the day.

Although Iraqi had abundant chemical weapons, they did not use them in Desert Storm. We interrogated their people as to why this was so. I thought perhaps it was a concern that we might retaliate with nuclear weapons, and certainly we made things ambiguous in that regard. But the Iraqi generals reported the reason they didn't use chemical weapons was because if they used them, they knew our troops were better protected than their troops. They felt that if they used chemical weapons, they would suffer many more casualties than we would.

I think that they were right in their assessment, and I think that it highlights the importance of defenses being a key element in deterrence. If you have ballistic missile defenses, then that makes another country's acquiring ballistic missiles less important, and in fact, may deter them from spending large amounts of money to build them.

Our allies and neutral nations would benefit greatly if we had wide-area ballistic-missile defenses. For example, suppose we had the capability to share ballistic missile defenses on a global basis, say a space-based system. You could go to a country such as Israel—which I believe has nuclear weapons—and say, "If you get rid of your nuclear weapons, we'll share ballistic missile defenses with you." Or suppose we went to India and Pakistan and said, "Look, we understand that you don't like each other, we understand that you are both relying on ballistic missiles and nuclear weapons to deter one another, and that if you have a war, there is liable to be a nonrational decision which causes one or the other or both to use these weapons. The problem is, we live downwind, and we don't like it. So, while you may want to use these systems, we're not going to let you. We are going to preclude you from doing that, at least to the extent we can, by using our ballistic missile defenses to interfere with any missile attack that takes place." I think that use of active defenses can be as important as using ballistic missile defenses in the direct defense of our own nation.

Consider this: The former head of the Japanese Air Force, General Ishizuka, and I have talked for hours about the challenges in defending against possible missile attacks, because it is a subject of very high concern in Japan. The military leadership in Japan is very concerned about ballistic-missile defense. One of their concerns is that their people, or elements of their population, will lose confidence in the United States as the protector of their security, and they are apprehensive about what effect that might have. If the Japanese decide they must go their own way, and renounce their constitution and develop a very aggressive outreach in military capability, what would that do to the whole Pacific Rim, or to the countries that well remember WWII? So, it's important that we have the ability to share ballistic missile defenses with a country like Japan.

Working with the Russians, the Japanese, and others will also affect the Chinese. The number of warheads the Chinese have is significant in terms of inflicting pain, but in terms of attacking Russia or the United States, they are probably not decisive. The Chinese are pragmatic, so you can work with them if you understand what their internal concerns are, such as control of the Chinese border. So I think that over the long term, if you ignore them in this regard and work with the Russians, the Japanese and countries like that, you can then say to the Chinese, "By the way, we're going to neutralize your ballistic missiles."

Does it mean we'll have peace? No, I don't think so. I think you have to ask yourself this one question: "Would it be a better world if we had ballistic missile defenses and no nuclear weapons, or nuclear weapons and no ballistic-missile defenses?" I think that this is the choice, and I think that is where we need to have the vision and the courage to take on this mission of ballistic missile defense in new-era warfare.

Reducing Nuclear Arsenals

One of the greatest potential benefits of ballistic missile defense—especially shared warning and wide-area defenses—is in reducing the value of nuclear ballistic missiles. There is a significant domestic value here as well, but the costs of developing, deploying, and operating such defenses are not trivial. In fact, in the current budget environment, you can't expect the military to take the lead. But I think ballistic missile defenses would compete very well, particularly if you trade them off against the cost of maintaining nuclear-weapons forces.

Our nuclear-weapons forces are not trivial in terms of cost. The Air Force has got to come to grips with relinquishing the ICBMs. We just put three black boxes on each one of the Minutemen, at a million dollars a box. The cost to the Navy operating the Trident submarines is certainly not trivial.

What are we going to have to do when we have to replace the tritium? Are we going to restart the Savannah River Project? So, I think if you worked a trade-off one against the other you could probably free up a great deal of money.

There is another argument made by those who cling to cold war ideas, and that is that we need to maintain our nuclear weapons to deter other nations. These people point to the proliferation of the technologies for weapons of mass destruction, particularly nuclear weapons, and worry that without our own nuclear arsenals we will be ripe for attack. But does our nuclear deterrent force really limit the threat posed by the proliferation of such weapons? Such weapons might be delivered by a variety of means. Aircraft and missiles carrying such weapons might not be the only threat. Those attempting to smuggle nuclear weapons into the United States might do so in a manner that is not traceable. But even if we identified the culprits, we would probably not be able to use a nuclear weapon against them. Nonetheless, we ought to have the conventional military power to be able to influence them on a deterrable basis.

In other words, the question is, *are we being well-served if, as we cut our military forces down, we reserve large amounts of money to preserve weapons we probably cannot employ, as opposed to acquring more very capable conventional forces?* If a nation uses a nuclear weapon or poison gas against us, then we must have military options to take—and strong conventional forces provide those options.

So, you have two problems: deterrence, and what do you do if deterrence fails. Anybody who would poison Chicago is nondeterrable, because they are not operating on our same logic train. Therefore, I think that some potential aggressors that are indifferent to whether we have nuclear weapons or not. I believe Saddam Hussein, in many ways, was indifferent to nuclear weapons. In fact, sometimes I wonder if he really was not looking for Israel to throw a nuclear weapon at him.

So, the idea of using US nuclear weapons to deter an adversary is probably only useful with reference to the Russians. That's what I believe.

However, let me be clear on this. We cannot—we must not—unilaterally disarm. You must do it *in concert* with the Russians, and you must have a vision. But we seem to lack vision, and to be in a reactive mode, which precludes long-term planning. What we mean by long-term is really a function of perception of threat, of building trust with Russia. I would not suggest that we can walk away from nuclear deterrence other than arm-in-arm with the Russians. There is an opportunity to let other nations have nuclear arsenals and yet work with them on a business-like basis. I personally am not threatened by England and France having nuclear weapons, or even Israel. I think China also tends to fall in that category. Nuclear disarmament has to be worked in terms of the Russia/US relationship.

Maintaining Conventional Strength

It is absolutely essential that we maintain US conventional military strength—the type of conventional strength that we exhibited in Desert Storm. I think that we are approaching what I call the dominance of defense. The thing that scares me, of course, is the French reached that conclusion in 1930. I see so little value in warfare.

For example, in Saddam Hussein's case, I am not sure that we missed the boat in not dethroning him. If I were somebody living in Baghdad, what threatens me most are the Kurds and the Shiites. So, while I might hate Saddam Hussein for what he does to my family, my economy, and my nation, an Iraqi might believe that he is the one person who was keeping me alive. If I were a Kuwaiti, I would hate the Iraqis for what they did to my children and my country, but on the other hand, I'm not sure I would want Iran on my northern border either.

Probably the mistake we made was in our publicity during the Gulf War, which demonized Saddam Hussein. He *is* a gangster and an evil person. But in the Vietnam War, we became very involved in the internal affairs of South Vietnam to our disadvantage, and I think the thing we were all concerned about was getting involved in solving the problems of Iraq, when Iraq's problems are not solvable. So, in my opinion, if the Iraqis want Saddam, they can elect him, let him try to run Iraq, but Saddam and Iraq would be *very* wise not to step over the boundary lines again.

Of course that's fine, if Saddam Hussein is just a modest risk taker. But if you are faced with a dictator who in fact runs risks, a defense-only capability may not be an adequate US military posture. We also should have a very credible conventional military capability.

I think we also have to remember that each war is different. We are ill-prepared for things like Bosnia or Somalia, and there are those who want to build forces for that. We fight a Desert Storm only every 10 or 15 years. But, I think the point is that you've got to be able to fight the Desert Storms successfully, and obviously counterforce is necessary. I'm not sure what the campaign against Iraq's economy achieved. But basically, our goal was to eject Iraq from Kuwait and cripple Iraq's nuclear/biological/chemical capabilities, and I do believe we did have an impact. We went after leadership, and we went after it in a military context, not in a political context. Since Saddam was the top Iraqi military leader, it would have been nice if we had gotten him, but only because he was their top military figure, not because he was their president.

Conclusions

The fundamental point is clear: the United States and its allies must pursue ballistic missile defenses. Ballistic missile defenses are the key to new era warfare—to opportunities for reducing nuclear arsenals and to ensuring the type of conventional strength that enables the United States to secure its interests in the future. As far as I am concerned, it is simply time to get on with it.

Chapter 3

The Revolution in Military Affairs

Jeffrey McKitrick, James Blackwell, Fred Littlepage, George Kraus, Richard Blanchfield and Dale Hill*

According to Andrew Marshall, director of the Office of Net Assessments in the Office of the Secretary of Defense, "a Revolution in Military Affairs (RMA) is a major change in the nature of warfare brought about by the innovative application of new technologies which, combined with dramatic changes in military doctrine and operational and organizational concepts, fundamentally alters the character and conduct of military operations." Such an RMA is now occurring, and those who understand it and take advantage of it will enjoy a decisive advantage on future battlefields.

Military theorists around the world have long noted the historical discontinuities in the conduct of warfare caused by the advent of new technologies and weapon systems. The Soviets called these discontinuities "military-technical revolutions." Recently, analysts in the United States have started calling them RMAs. This change in terminology was meant to capture the nontechnical dimensions of military organizations and operations, the sum of which provide a large part of overall military capabilities.

The nature of these discontinuities is such that warfare after the "revolution" is unlike what went on before in profound and significant ways. Throughout history, there have been a number of such revolutions. Gunpowder produced an early military revolution in the Western World, transforming both land and naval warfare. During the mid-nineteenth century, industrialization revolutionized warfare through railroads, the telegraph, the steam engine, rifled guns, and ironclad ships. More recently, the mechanization of warfare during the

---

*Robert Kim, Mark Jacobson, John Moyle, and Steven Kenney also assisted in the preparation of this chapter.

interwar period led to the development of blitzkrieg, carrier aviation, amphibious warfare, and strategic bombing.

In some cases, the changes in technology associated with these revolutions changed not only transportation, communication, and warfare but also entire societies as well. During the transportation revolution, for example, railroads altered the economies of nations and allowed them to move military forces farther and faster and sustained them longer. Moreover, these societal changes created new sets of operational and strategic targets. We currently characterize these kinds of revolutions as "social-military revolutions."

To date, the bulk of the intellectual and physical development associated with the current RMA has focused on new systems and technologies. What is needed now is a more careful analysis of the new operational concepts and new organizations that might best help us realize the full potential of these new systems and technologies. To reach that level of analysis, we need to start with an appreciation of the historical and geostrategic contexts in which the RMA may unfold.

What motivated past changes in the conduct of warfare? Who might our future competitors be? What will be their political and military objectives? How might they choose to organize and equip militarily to achieve those objectives? How might the conduct of warfare change? The answers to these questions will assist us in identifying new RMA warfare areas and, in turn, help identify what new military capabilities the United States will need.

Before proceeding, however, we must issue a word of caution. Although we think that we now stand at the start of a long period in which we may face a RMA, we cannot be certain about when the transition period might start, how long it might last, what new competitors might arise, when they will arise, or what new warfare areas might be developed, not to mention a host of other key questions. In short, we do not have an absolute grasp of the scope, pace, and implications of this possible RMA.

We can make useful observations even at this early stage. During and immediately after the First World War, forward-thinking military officers such as Col J.F.C. Fuller of the British army and Maj Earl Ellis of the US Marine Corps outlined the basic features of armored warfare and amphibious warfare. They defined these concepts decades before the necessary systems existed and at a time when the political circumstances of the next war were uncertain.

More recently, science fiction writers have been exploring future war-fighting capabilities. The novels *Starship Troopers* by Robert Heinlein (1957) and *Ender's Game* by Orson Scott Card (1977) alluded to military systems that today equate to artificial intelligence and virtual reality. These future systems appeared to be pure fantasy at the time, but yesterday's fiction has, in many instances, become today's reality. It should be instructive to those of us engaged in looking to the twenty-first century that military officers and authors were able to look into the future and picture types of warfare that became real or today are becoming real.

Achieving analytic progress requires discussion of these issues with as much definition and certainty as possible in order to give analysts something concrete to critique and, thereby, improve. It is our hope that such discussion will facilitate identifying new ideas and animate a discovery process that includes war games and simulations over the course of the next three to five years. It is in this spirit of discovery that we offer the following portrayal of the unfolding revolution in military affairs.

Lessons of Past RMAs

RMAs have risen from various sources, with many—but not all—of them technological. Societal change contributed to a military revolution during the wars of the French Revolution and the Napoleonic era, in which the *levée en masse* allowed for the creation of larger, national armies.

In the technologically based military revolutions of this century, different scientific fields have provided the enabling factor. For example, chemistry and early physics drove many of the critical advances during World War I. In this war of gunpowder, the rate at which weapons fired and the ranges that the projectiles traveled decided the fate of many battles.

Advanced physics drove the next RMA, which extended from the mastery of flight to improved radios and the introduction of radar through the creation of nuclear weapons at the end of World War II.

The current RMA has as its source what has been called new physical principles. These principles focus on technologies such as lasers and particle beams. Current trends indicate that the next revolution in military affairs may have a biological source. Some manifestations of these biological advances may include biosensors, bioelectronics, nanotechnologies, distributed systems, neural networks, and performance-enhancing drugs.

New technologies and systems significantly influence the RMA, although the resulting RMA could take one of a number of forms. The interwar innovations of armored warfare by the German army, amphibious warfare by the US Marine Corps, carrier warfare by the US Navy, and strategic bombing by the US Army Air Forces have been characterized as "combined-system RMAs." Their revolutionary nature derived from a collection of military systems put together in new ways to achieve a revolutionary effect.

A different type of RMA is the "single-system RMA." An example is the nuclear revolution of the 1940s and 1950s, in which a single technology, nuclear fission/ fusion, drove the revolution. Another example of a single-system RMA is the gunpowder revolution, in which gunpowder transformed land and naval warfare through the use of siege guns, field artillery, infantry firearms, and naval artillery.

Evidence suggests that the revolution unfolding today is neither a combined-system nor a single-system RMA but an integrated-system RMA. The outlook is for the rapid evolution of new technologies eventually leading to the development of several advanced military systems.

These systems, when joined with their accompanying operational and organizational concepts, will become integrated systems. In contrast to developments during the interwar period, this system-of-systems approach will aim to take advantage of the cumulative effect of employing each of the new capabilities at the same time.

In World War II, each new form of warfare took place in its own operating medium—armored warfare on land battlefields, strategic bombing in the air over homelands, carrier warfare at sea, and amphibious warfare at the intersection of land and sea—and only occasionally interacted with the others. In the current RMA, the integrated employment of all the new systems will be essential to take advantage of their true value.

Nevertheless, this forecast does not exclude the possibility of a single-system RMA. To avoid strategic surprise, we must continue to think about breakthroughs in critical areas such as information technology, biogenetics, and others. Unforeseen advances in these areas could bring about a sudden, significant, and solely owned military advantage to the country that achieves a breakthrough.

The same holds true for a combined system RMA. Furthermore, there is also the possibility that the information revolution may result in far-reaching societal changes, putting us on the path of a social-military revolution. Such a revolution holds profound, but somewhat different, implications for the changing nature of warfare.

It is important to remember that technologies and systems enable but do not cause military revolutions. Past military revolutions have been driven by requirements that have motivated military organizations to innovate in order to overcome the limitations of existing practice. Such strategic, operational, and tactical requirements determined whether technologies were adopted and how they were employed. Without them, stagnation can prevail, even in states possessing technologies with revolutionary implications.

An example of this principle is the gunpowder revolution of the sixteenth and seventeenth centuries. In Europe, gunpowder weapons fundamentally changed the conduct of all areas of warfare—maneuver warfare, siege warfare, and naval warfare—on account of the constant competition between rival states of roughly equal military power. Imperial China developed gunpowder and firearms a century before Europe possessed them, but stagnated in all areas. China fell behind in the gunpowder revolution largely because of its vast population, which allowed it to overcome any land or sea threats through sheer weight of numbers.

In contrast, the smaller states of Asia made significant innovations in response to pressing military requirements. In sixteenth century Japan, where rival warlords strove for dominance, the rise of firearm-equipped infantry and the use of volley fire mirrored developments in Europe. Korea in the 1590s responded to the threat from reunified Japan by developing its "turtle ships," ironclad, cannon-armed galleys that provided a technological advantage that proved essential to defeating three successive Japanese invasions.

Another example is the interwar period, in which all of the major powers possessed the same technologies but only a few countries created new operations concepts and organizations. For example, the development of carrier aviation took place in the United States and Japan for the purpose of fighting major naval engagements in the Pacific. Carrier aviation withered in Britain's Royal Navy, whose main tasks were fighting in confined waters such as the Mediterranean Sea and combating German commerce raiders on the high seas.

In the case of armored warfare, only the Germans employed tanks, radio, and airplanes in new ways in 1940 even though all of the essential technologies had been available since World War I. This situation was due in part to Germany's unique strategic problem of being surrounded by enemies. This problem led to an operational requirement for rapid offensives to defeat enemy states quickly, and the emphasis on offense dictated tactical requirements for mobility, firepower, and protection.

Germany thus integrated tanks, infantry, and artillery in the Panzer division and supplemented them with close air support. France and Britain, constrained by defensive strategic and operational concepts on land, did not innovate and paid the price in May of 1940.

Past RMAs hold another significant lesson for the current RMA. In past RMAs many of the key systems were already used in combat or in civilian applications decades before significant changes occurred in military organizations. For example, railroads began carrying commerce in the 1830s, but in the 1860s only the Prussian army under Helmuth von Moltke, the elder, used them to facilitate intricate mobilization plans that conferred a significant operational advantage at the start of a campaign.

Similarly, tanks, radios, and close support aircraft were used in quantity in World War I, but they did not realize their true potential until the Germans devised new organizational and operational concepts for them in the 1930s. Likewise, the implementation of revolutionary operational and organizational concepts in this RMA may require a long time even though most of the key systems probably are already in development or have even been used in combat. We need to start thinking immediately about the shape of warfare in 2020 in order to capitalize on the RMA in a timely fashion.

Future Competitors

If we hope to understand the scope and potential impact of the RMA, we first must understand our potential future competitors. The threats and the vulnerabilities of these competitors will clearly influence how the United States should exploit the RMA.

We must stress that we are talking about potential *military* competitors—not political and economic competitors—although politics and economics are related factors. Developing and assessing alternative futures—projecting the nature of future competitors, their force structures, and modes of operation—is far from an exact science. Indeed, if the events of the last five years are any indicator, current approaches at predicting the future will continue to meet with only marginal success.

While it is difficult, if not impossible, to assess accurately which nations will become peer competitors in 25 years, it is possible to imagine one nation, or a combination of nations, rising to challenge US national security interests. History provides many examples of potential competitors rapidly elevating themselves first to regional and then to peer competitor status. Japan and Germany prior to World War II are examples of the ability of nations to evolve rapidly up the competitor scale through willpower and sacrifice. To assume that we are unlikely to see a peer competitor in the next 25 years is to ignore history.

Given that a peer competitor is likely to emerge, the most important question is not who the competitor is but the likely characteristics of that competitor. Leading experts in the field of assessing the nature of future competition have provided significant insights into how we view and think about potential competitors.

Dr Paul Bracken of Yale University, in his article on "The Military After Next,"[1], characterizes nations as Type A, B, and C competitors. Type A competitors are peer competitors, able to compete with the United States on a global basis across a full range of military capabilities. Type B competitors are regional competitors, able to compete regionally, and only across a limited set of military capabilities. Bracken's Type C competitors are terrorists, low-intensity conflict countries, drug lords, and the like. We feel this type of competitor is not really a national security competitor but a political competitor. It is more useful to think of Type C competitors as being niche competitors. A niche competitor would be a country that has chosen to specialize in a specific military capability that appears to have high leverage against US forces. This type of characterization seems useful in bounding the range of possible competitiveness in a military sense.

Dr Stephen Rosen of Harvard University has framed the debate on future competitors by dividing the world into "zones of peace" and "zones of turmoil."[2] His fundamental division of the great powers places the industrialized democracies in the zone of peace with all other countries in the zone of turmoil. Conflicts will most likely arise among nations in the zone of turmoil, or between them and countries in the zone of peace. It is difficult to develop scenarios that lead to wars among nations in the zone of peace.

So long as the countries remain democratic, Dr Rosen cannot envision a war between the United States and, for example, a democratic Japan or Germany. Dr Rosen's analysis reminds us that it is important to consider political traditions, cultural norms, economic strength, alliances, and many other factors as we think about the evolution of competition between states. Furthermore, states may swing between peace and turmoil, dramatically change goals and directions, or gain or lose military capabilities, making our strategic planning even more difficult.

To understand the nature of future competitors and, more importantly, plan for their potential emergence requires an understanding of the heart and soul (or national character) of the countries under consideration. By "national character" we mean what makes them tick—what factors (current and future) would push them along the path from niche, to regional, to peer competitor.

Such considerations might include, but are not limited to, the following:

- Historical context
- Cultural and social beliefs (mores, etc.)
- Demographics (rate of change, age of population)
- Geography (landlocked, access to ports, agriculture)
- Economic development (trade, industrial base—indigenous/other)
- Political system (democratic, autocratic, stability)
- Access to foreign markets and technologies (sunrise/sunset systems, information technologies and capability)
- Military force structure (disposition, training)
- Nature of alliances (cooperative agreements, traditional/nontraditional adversaries)

In other words, it is essential that we understand what a nation or region is to more accurately reflect what it may become.

Understanding the national character and proclivities of a nation is only part of the equation. Equally important is an appreciation of national, regional, and/or global trends that may act in synergy with national character to propel a nation to competitive status. Trends associated with economics, the pace of technological innovation, the development of military weapon systems, the growth of new operational and organizational approaches, and the proliferation and diffusion of military systems and technologies will all interact with the national character and daily events of virtually every nation in the world.

While some analysts may view the birth of competitor states (China or Japan for instance), as being quite predictable, we should be cautious in ascribing too much certainty to such predictions. Trends may cause nations to transform in unanticipated ways, thereby giving rise to a number of surprise competitors. Nevertheless, evaluating trends in the context of national character may help to narrow the field of who these competitor states may be.

Taking into account our understanding of the national character of potential competitor states/regions, our subsequent analysis should focus on the three dimensions of the RMA required for a nation to achieve competitor status. First is the conscious decision on the part of a state to acquire all or portions of what might be termed an RMA complex. Second is the ability to acquire or develop the systems that constitute RMA-type technologies. Last, and perhaps most important, is the ability, organizationally and operationally, to adapt technologies in ways that bring into being the full military potential of an RMA.

In the analysis of the current RMA, which may take decades to emerge, analysis of likely competitors will be a long, arduous, and ongoing process. Today, we could identify a number of potential candidates as peer competitors (Russia, Japan, China, Germany, a unified Korea, an Asian coalition, etc.), but whether they achieve that status remains in question. An even more interesting result of the analysis may be the surprises—new candidates, not unlike Japan in 1940, which have the potential to rise from niche to regional or peer competitor status in the early decades of the twenty-first century.

In either of these eventualities, it is likely that for a period of time in the future, the United States will be faced with a number of regional competitors while not being forced to deal with a true peer competitor. However, historical precedent leads us to believe in the inevitability that a large peer competitor will emerge over time. It will be critical for the United States to understand the nature of the challenges posed by such a peer long before the competitor achieves rough parity across important military areas.

Understanding the natures of competitors is especially important because of their influence on emerging forms of warfare. Judging from the example of past RMAs, distinct approaches to harnessing new technologies will surface in those states seeking to exploit them. During the interwar period, the Germans, British, and French followed different directions in the usage of tanks; similarly, the US and British navies differed in their development of carrier aviation.

The most successful approaches will not necessarily come from the countries most experienced with the relevant technologies, as evidenced by Britain's lagging in the development of tanks and carrier aviation despite having invented them. Success will most likely come through favorable matchups between competing doctrines. Obtaining an advantage over a peer competitor through the RMA will require understanding likely opponents' tendencies in order to determine the optimum approach.

New Warfare Areas

We think the current RMA, like the interwar period, will involve the emergence of multiple new warfare areas. A warfare area is a form of warfare with unique military objectives and is characterized by association with particular forces or systems. Examples of warfare areas that emerged in the interwar period are armored warfare, carrier warfare, amphibious warfare, and strategic bombing. We have currently identified four potential new warfare areas—long-range precision strike, information warfare, dominating maneuver, and space warfare. Other areas that we have yet to identify could also develop.

Of the four potential new warfare areas, precision strike is the most developed conceptually, although even here much analytic work remains to be done. Much work has been done in the area of information warfare, yet it remains a poorly understood concept. Analysis of dominating maneuver and space warfare has just begun.

The warfare areas that we have identified are likely to emerge in the long run but will not necessarily be developed fully in the near future. Doctrinal development is a long and uncertain process, and military history offers numerous examples of unexploited warfare areas—concepts intended to revolutionize warfare that did not come to fruition. Technological limitations, conflicts with prevailing doctrine, or lack of strategic purpose derailed these developments.

In the late nineteenth century, the *Jeune École* in France sought to exploit an emerging weapon, the torpedo, to contest British sea control with small, cheap torpedo boats for commerce raiding and coastal defense. Their attempt to create a new warfare area led to a decade of doctrinal uncertainty in naval warfare but failed by the turn of the century owing to the ineffectiveness of the primitive torpedoes and torpedo boats, the rising influence of Adm Alfred Thayer Mahan, and the emergence of the Anglo-French entente.

After the Korean War, Gen James Gavin and others in the US Army sought to create a new form of land warfare using the helicopter. Seeking greater strategic mobility between theaters and a "mobility differential" over the battlefield, they envisioned helicopter-equipped units that could rapidly deploy in a crisis and would use their superior mobility to their advantage in the cavalry roles of scouting, pursuit, and delaying actions. They succeeded in making helicopter aviation a significant part of the Army, but the Army developed the helicopter as part of a combined-arms team rather than as the basis for autonomous units, fielding only one air assault division during and since the Vietnam War. The vulnerability of helicopters to air defenses and the predominance of armor and infantry in existing doctrine each contributed to this result.

Short-term technological and doctrinal barriers will not diminish the ultimate importance of a new warfare area. There are past examples of warfare-area concepts that were abortive in one context but resurfaced in other settings with the emergence of the right enabling technologies or doctrinal pressures. The ideas of the *Jeune École* appeared again in Germany during the First World War, when practical submarines were the enabling technology and the need to strangle British commerce provided the doctrinal pressure.

Similarly, the nascent armored warfare concepts of J.F.C. Fuller and the Salisbury maneuvers went undeveloped in Britain but reemerged in the German army. The fact that concepts discarded by Britain and France provided the basis for U-boat warfare and the Panzer divisions illustrates that fundamentally sound concepts will eventually be exploited— if not by the United States, then by its competitors.

Precision Strike

Precision strike may well be the most thoroughly understood new warfare area of the next revolution in military affairs. This is true because the United States has been a leader in the development and deployment of such systems since the 1970s. The creation of precision strike capabilities during the latter stages of the Cold War in fact cast a long technology shadow and deeply affected Soviet military thought.

The Red Army gave us far more credit than we deserved for developing the reconnaissance strike complex. It probably saw tests of parts of systems, such as the Defense Advanced Research Projects Agency (DARPA) "assault breaker" missiles in the mid-1970s, as indicative of imminent deployment. Somewhat ironically, the Soviet military may have more fully comprehended the revolutionary impact that such systems would have on future battle than did the US military establishment.

Precision strike in the context of the coming RMA is well beyond its predecessors of follow-on forces attack (FOFA) and joint precision interdiction, which are its conceptual forebears. At the time of the development of such precision strike systems as the Joint Surveillance Target Attack Radar System (JSTARS) and the Army Tactical Missile System (ATACMS), the idea was to create a maneuver differential for NATO ground forces.

NATO planned to delay and disrupt the arrival of second-echelon and third-echelon Warsaw Pact armored forces before they could overwhelm the outnumbered and outgunned NATO defenders. Such deep strikes would extend the battlefield; they would delay in time the advance of the conflict to the nuclear threshold; that would permit reinforcing forces to arrive from the continental United States (CONUS). This was planned to allow for the creation of a conventional counteroffensive force to turn back the attacking Warsaw Pact armored advance.

The Gulf War demonstrated the potential for such deep strike systems not only to create a maneuver differential, but at least potentially to be decisive in themselves. Precision strike, in the context of the unfolding RMA, is the ability to locate high-value, time-sensitive fixed and mobile targets; to destroy them with a high degree of confidence; and to accomplish this within operationally and strategically significant time lines while minimizing collateral damage, friendly fire casualties, and enemy counterstrikes. In 2020, precision strike technologies will create the potential to achieve strategic effects at intercontinental distances.

The potential effect of precision strike can be seen in the dramatic increase of capabilities to strike strategic targets. In 1943 the US Eighth Air Force prosecuted only 50 strategic targets during the course of the entire year. In the first 24 hours of Desert Storm, the combined air forces prosecuted 150 strategic targets—a thousand-fold increase over 1943 capabilities. By the year 2020, it is not out of the realm of possibility that as many as 500 strategically important targets could be struck in the *first minute* of the campaign—representing a five thousand-fold increase over Desert Storm capabilities.

It is envisioned that precision strike will be able to achieve effects similar to those of nuclear weapons but without the attendant risk of escalation to intolerable levels of destruction. When directed against targets comprising the enemy center of gravity, precision strike might itself prove decisive. For it to do so, however, requires a much clearer understanding than we have had in past wars about what constitutes the enemy's center of gravity and what it takes to affect it.

The precision strike area of warfare presents a significant challenge to the organizational adaptation of the US military. These systems achieve decisive impact only if they are integrated at the operational or strategic level of war. This means that a single theater or global commander must have control over the employment of precision strike systems, as was done during Desert Storm.

A potential problem does arise however in the development of current precision strike systems, which are jealously guarded by the individual military services and even by some DOD-wide agencies. This arrangement tends to lead to the acquisition of systems that are duplicative rather than complementary. Notwithstanding the potential benefits to be derived from having the services compete for the precision strike mission area, there is a potential danger that in the current budgetary climate such competition may prove to be counterproductive.

There is a further potential problem in that even should we manage to orchestrate a successful joint program management scheme, the architecture that frames the development of disparate precision strike technologies and systems may not be structurally coherent. This could result in several separate and perhaps inherently incompatible components, none of which have much military utility by themselves. There is a growing need for a system-of-systems framework to define precision strike requirements, as well as a need for an architecture which would drive the development of advanced technologies and systems applicable to precision strike.

We must begin now to move beyond the service-specific approach to the employment of precision strike and experiment with new organizational approaches to employing these systems on the battlefields of 2020. These systems will likely be theater-wide or even global, and there will be no single service able to provide a service-specific core organizational unit to serve as the basis for the commander-in-chief's joint task force for precision strike operations. Therefore, new organizational units may have to be cut out of whole cloth.

This does not imply that massive new commands or new organizations must be created. In fact, we may need to consider just the opposite approach. We may need to consider the power inherent in the new information technologies that could greatly expand a commander's span of control, allowing us to eliminate one or more levels of command and to consequently accelerate the decision-making and command and control cycle.

The essence of precision strike is the ability to sense the enemy at operational and strategic depth, recognize his operational concept and strategic plan, and select and prioritize attacks on enemy targets of value. All of this is intended to achieve decisive impact on the outcome of the campaign. To be most effective those attacks probably should be synchronized in time and space.

The revolutionary potential of precision strike derives from the technologies that provided a glimpse of their own potential during Operation Desert Storm. These and related technologies enable commanders to have continuous wide-area surveillance and target acquisition, near-real-time responsiveness, and highly accurate, long-range weapons at their disposal.

Such technologies by themselves have the potential to change dramatically the way wars are waged. Integrating precision strike capabilities with dominating maneuver and information war may create an especially potent RMA.

By 2020, real-time responsiveness of sensor-to-shooter systems must become a reality. For the first time in history, this responsiveness will allow the striking force to maneuver fires rather than forces over long ranges, and allow direct and simultaneous attack on many of the enemy's centers of gravity.

This warfare area is currently being driven by advances in technology. The key improvements that are now occurring are in broadening the environmental conditions for wide area surveillance and precision targeting; security and countermeasures; data processing and communications; delivery platforms; precision munitions; and positioning/locating devices. The advances needed to exploit our lead in this warfare area include continuous situation awareness and improvements in data fusion, mission planning, and battle damage assessment (BDA).

At the same time, we need an equal effort in developing new operational concepts and organizations for the application of precision strike. As in other new warfare areas, it may be that the greatest military payoff will come from operational approaches and organizational adaptation—not from systems.

Information Warfare

Another revolution under way in warfare is that associated with information systems, their associated capabilities, and their effects on military organizations and operations. We call this new warfare area information warfare, which we define as the struggle between two or more opponents for control of the information battlespace.

At the national level, information warfare could be viewed as a new form of strategic warfare, one of the key issues being the vulnerability of socio-economic systems, and the question is how to attack the enemy's system while protecting yours. At the military operational level, information warfare may contribute to major changes in the conduct of warfare; therefore, one of the key issues is the vulnerability of command, control, communications, and intelligence systems, and the question is how to attack the enemy's system while protecting yours.

As we increasingly assimilate information capabilities into our military structure and focus more and more on establishing and maintaining an "information advantage" as a war-winning strategy, we also change the vulnerabilities of US forces, and, ultimately of the United States itself. The force structure that will implement information warfare 25 years from now may well be different from today's military in more ways than just its equipment. Moreover, the character of warfare may change in ways that affect our thinking regarding intelligence and crisis and wartime decision making.

Some of the changes might include the whole issue of deciding that a war has begun. It is not clear at this time whether information warfare measures taken by a potential adversary at the outset of a war would be readily detectable. The question of how you know you are at war may be difficult to resolve in view of the potential ambiguity associated with information warfare.

Ambiguity and plausible denial are not new phenomena. But the rapid growth of interconnections manifested already in communications, banking, and other areas creates vulnerabilities and presents opportunities to do grievous harm—quickly, with no warning, and with a minimal "signature." Accordingly, the analysis of indications and warning that mark the outset of warfare must change. To date, insufficient thought has been applied to this aspect of the character of future war.

In addition to its inherent ambiguity, information warfare in 2020 also portends a very different set of potential responses by the United States to an adversary detected acting in a hostile or potentially hostile fashion. The information warfare measures that the United States could take might require quite different policies with regard to rules of engagement than have previously been contemplated.

This is both good news and bad news. The good news is that there may be new tools, short of lethal attacks available to signal an adversary that warfare with the United States would be a bad idea. The bad news is the inherent ambiguity noted above. In other words, it might be hard to signal our intentions. In any case, the point is that much thought must be given to the examination of threats, requirements for intelligence, and rules of engagement.

Although countering an adversary's command and control has always been a feature of warfare, the continental United States has been somewhat invulnerable to such measures. One clear implication of warfare in 2020 is that almost any enemy will try to degrade our information system. The United States must be prepared for that eventuality. Paradoxically, although the technology of information systems is becoming more capable and sophisticated, it is actually harder to secure the US information infrastructure from attacks.

One potential vulnerability is the fact that information warfare generates problems at the national level rather than just for the Department of Defense. Therefore, such problems will not be solved by creating new military organizations. The problem goes beyond the armed forces to the entire national security infrastructure. As the international information infrastructure grows and elaborates, its reach expands beyond the control of any single entity or any single nation.

Thus, the infrastructure is beyond the control of those who use it, and has access points at a myriad of places for others to enter the system. It is not at all clear, as a result, that today's military organizational structure (e.g., the Joint Chiefs of Staff, regional and supporting commanders-in-chief, and military service departments) is the best way to manage the complexities of information warfare as it might unfold in 2020.

Another key organizational deficiency is the lack of a coherent strategic approach to offensive and defensive considerations for information warfare. Nowhere in our system do we bring together and integrate the offensive and defensive nature of information warfare. Nor is there a single locus for requirements generation and staking claims in the acquisition process for information warfare.

The Defense Information Security Agency (DISA) has been actively pursuing "information assurance" as a part of the charter to take responsibility for all DOD information systems. However, DISA is not an operational command.

The question is, will we be able to plan for war in 2020 with the view that communications channels will be available, with little concern for the overall architecture or the nature and characteristics of those channels? It seems clear that information warfare in 2020 may require operators to have more familiarity with how commands interact in order for these operators to execute effectively strategy and operations.

There are examples of the problems we can anticipate if such "information awareness" is not part of the force. The first example concerns the logistics information systems, which are both elaborate and critical to successful military operations. Despite their criticality, these systems are generally subject to less stringent security measures than other military systems. The shortcoming does not involve just cryptologic security, although that is an area that lags operational channels. We could make the system cryptologically secure but still be quite vulnerable. A potential enemy could significantly disrupt our operations merely by denying us information—by simply interfering with logistics transmission links.

A second example is the anticipated problems associated with the presence of automated systems on the battlefield of the future. These systems are likely to be considerably more widespread in 2020 than they are today. The commander who knows his "human" systems but who does not understand his "automated" systems will be vulnerable to surprise—possibly to defeat. Knowing the automated systems entails an understanding of the software programming "rules."

There are already many autonomous systems available to the commander to include unmanned air vehicles (UAVs), unmanned underwater vehicles (UUVs), smart mines, and Tomahawk cruise missiles, among others. The commander who does not understand the details of his logistics information system, or the programming of his autonomous systems, may face a significant information warfare vulnerability in 2020.

In dealing with the information revolution that is affecting the military today, the US military services seem to be engaged in improving their current communication channels. That is, they are striving to improve performance elements within the current organizational structure. They have yet to address the implications of systems and capabilities that do not fit within the current structure. This is a fundamental issue. The military traditionally has viewed information services, including intelligence and communications, as supporting inputs to the actual warfare functions of fire, maneuver, strike, and the like.

However, information warfare might not always be a supporting function; it might take a leading role in future campaigns. This makes it both more important and more challenging to get the organizational issue right. By 2020, at least in some militaries, the requirements of the battlefield will be such that traditional hierarchical command and control arrangements will be obsolete. In most organizations today, the decentralization trend is already well established.

Information technology is making distributed systems commonplace, and "virtual organizations" are growing like cultures on a petri dish. The rapid rate of growth of these types of new organizational entities would seem to suggest strengths that the military would be wise to examine.

Dominating Maneuver

One of the more recently identified potential new warfare areas is dominating maneuver. Maneuver has always been an essential element in warfare, but the RMA potentially offers the ability to conduct maneuver on a global scale, on a much-compressed time scale, and with greatly reduced forces.

We define dominating maneuver as the positioning of forces—integrated with precision strike, space warfare, and information war operations—to attack decisive points, defeat the enemy center of gravity, and accomplish campaign or war objectives.

While precision strike and information warfare are destroying enemy assets and disrupting his situational awareness, dominating maneuver will strike at the enemy center of gravity to put him in an untenable position, leaving him with no choice but to accept defeat or accede to the demands placed on him.

War is typically nonlinear, meaning that the smallest effects can have unpredicted, disproportionate consequences. In meteorology, nonlinearity is illustrated through the "butterfly effect"—a butterfly flapping its wings in the southern hemisphere can set off a string of reactions that eventually result in a violent storm in the northern hemisphere.

In the early nineteenth century, Clausewitz made similar observations when discussing the formulation of successful strategy. He wrote that victory comes not through winning battles or inflicting attrition but through attacking the enemy center of gravity, which—depending on the situation—could be his army, his capital, his leaders, or his principal ally. In the course of the twentieth century, it appears that the complexity of warfare has increased as military forces and their logistical and political underpinnings have become more complicated.

With increasing complexity, the nonlinear nature of war is likely to increase. Dominating maneuver seeks to exploit the increasing complexity and nonlinearity in warfare by striking directly at the enemy center of gravity in order to disrupt his cohesion and cause his swift collapse.

Dominating maneuver is distinct from maneuver in several ways. Maneuver refers to the "employment of forces on the battlefield through movement in combination with fires, to achieve a position of advantage with respect to the enemy in order to accomplish the mission."[3]

Dominating maneuver refers to the positioning of forces, not necessarily their employment; they can be positioned anywhere in a theater, not necessarily on the battlefield. It goes beyond "combination with fires" by integrating its effects with the effects from precision strike, space warfare, and information warfare. Its ultimate purpose is directly to achieve campaign and war objectives, transcending the role of ordinary maneuver.

Dominating maneuver does not require superiority at all points in the battlespace or imply domination of the entire maneuver. By 2020 our competitors could well challenge our national interests in regions where they enjoy the advantage of close proximity, and we may have neither a lengthy buildup period to marshal our forces nor access to a continental infrastructure to support our forces in the theater.

Under these circumstances, we will have to fight using the continental United States as our principal base of operations, making the maneuver battlespace orders of magnitude larger than it was in Desert Storm. Dominating maneuver could allow ground forces to operate successfully in situations where they cannot dominate the entire battlespace. The concept has a number of other implications for operations, organization, and technologies as well.

First of all, dominating maneuver will require new operational concepts that take into account the decisive importance of time, making future maneuver more simultaneous than sequential. It will be essential to attain operational and strategic objectives through simultaneous information warfare, space warfare, precision strike, and maneuvers against the enemy's critical points rather than through a series of pitched battles against enemy forces.

Further evolution of the Army's airland operations or the Marine Corps's operational maneuver from the sea could lead to such a concept. On the other hand, entirely new concepts may be required.

The German invasion of Norway in April 1940 was a campaign waged successfully in a way analogous to dominating maneuver, although on a smaller scale. A single airborne maneuver into Oslo on April 9, 1940, induced the surrender of the city's garrison by creating the perception that their cause was hopeless. Fighting continued on the fringes of Norway for six weeks, but the airborne maneuver led directly to decisive results. The maneuver gave the German commander a time advantage during which he could reinforce faster than the Norwegians could mobilize or the Allies could deploy. Moreover, the surrender of the Oslo garrison precipitated the capitulation of the Norwegian monarchy and forced the Allied decision not to become heavily engaged on the Scandinavian peninsula.

The Inchon landing during the Korean War was another operation that illustrates the principles underlying dominating maneuver. Gen Douglas MacArthur's plan to capture Seoul through an amphibious landing at Inchon struck at one of the critical vulnerabilities of the North Korean forces—their dependence on the transportation bottleneck at Seoul. Instead of gradually rolling the North Koreans back from Pusan, MacArthur planned to cut them off and put them into an extremely vulnerable position. The landing paralyzed the already overstretched North Korean forces, and they broke into disorganized fragments that retreated in disarray, incapable of serious resistance.

To execute dominating maneuver in 2020, the United States will have to develop new means for the movement of ground forces. The development of forms of mobility not possessed by the enemy could help generate maneuver dominance. More advanced concepts for comparable forms of mobility also could give a decisive advantage, as the Germans demonstrated in May 1940. The Germans generated maneuver dominance on the ground by employing combined-arms units that could mass combat power quickly, supporting them with fast-moving "aerial artillery" in the form of close air support, stressing aggressiveness, developing a faster command and control system to establish C^2 dominance, and employing air interdiction to degrade Allied mobility.

The revolutionary period that we are currently entering will have its own forms of maneuver that will require new technologies, new operational concepts, and new organizations. Along with new forms of mobility, the United States will require advances in logistical support to maintain the effectiveness of forces engaged in dominating maneuver. The need to operate far from existing bases, with little time for logistical buildups and with insecure lines of communications, will compel changes in the supply and support of ground forces.

The dangers of failure in this area are exemplified by the German campaign in the Soviet Union in 1941. In this campaign, the Germans quickly established complete maneuver dominance over the Soviets, whose decimated air force and poorly organized ground forces could not stop the Germans' fast-moving armored columns.

The Germans intended to use their maneuver dominance first to attrit the Red Army and then to seize Moscow, whose capture would sever the Soviets' transportation network and paralyze their political apparatus. Despite their maneuver dominance, the Germans failed—largely because of their logistics, which were inadequate to the needs of supporting far-flung advances over the vast distances of the Soviet Union. Future attempts at dominating maneuver may similarly come to grief as a result of shortcomings in logistics.

The dual imperatives of mobility and logistics may create the need for smaller forces and new transportation technologies. The current Strategic Mobility Study has as its objective the intercontinental deployment of a heavy brigade in 15 days and a heavy corps in 75 days. In the future, the United States may require the ability to move a corps-equivalent force across the oceans in seven days or less.

This capability might be achieved through the exploitation of new transportation technology such as fast sea transports capable of 100 knots or more, the national aerospace plane, and supersonic transports; through organizational changes creating smaller units with useful combat power that could deploy faster or be forward-deployed on naval platforms; or through ways not yet conceived.

The organization and tactics of such ground forces are difficult to visualize today. Some have suggested that twenty-first century variants of the so-called Hutier tactic developed by the Germans in World War I—Stingray or infestation tactics—would be useful. Such tactics would combine deception and bombardment with infiltration and attacks against strong points. The ground forces may be a small number of Army infantrymen, marines, or special operations forces, delivered deep in enemy territory by air and equipped with high-technology linkages to space-based or atmospheric strike systems, in effect acting as part of a sensor-shooter network.

The United States may need new technologies if it employs such tactics and seeks to maintain the lead that its forces possess in close combat. As advanced sensors and conventional weapons technologies proliferate and provide greater stand-off ranges for enemy forces, the United States should concentrate on achieving capabilities that will allow it to leap ahead of these developments. We should begin now to apply low-observability techniques to maneuver systems. We need to develop advanced propulsion technologies to give our maneuver systems greater speed, range, and agility. We also need new means to enhance the lethality of our munitions and the protective characteristics of our materials and systems.

Progress in these conceptual and technological areas will enable maneuver to play a significant role in the RMA. It is possible that precision strike and information warfare will make maneuver unnecessary in certain situations or that enemy progress in these areas will make maneuver difficult. Nevertheless, maneuver will be essential against an enemy unwilling to concede defeat unless the United States defeats centers of gravity that cannot be attacked without maneuver forces. Dominating maneuver may provide the coup de grace in future wars, and in other situations may serve as the enabler for war-winning space warfare, information war, or precision strike operations.

Space Warfare

We define space warfare, the fourth future warfare area of importance, as the exploitation of the space environment to conduct full-spectrum, near-real-time, global military operations. It includes facets of the other three warfare areas but has the potential to become a qualitatively distinct warfare area in its own right.

The US military's increasing reliance on support from space-based systems for its everyday operations and especially during times of conflict has highlighted the importance of space operations. However, space assets could provide more than support for the terrestrial war fighter in the future. The space environment offers the possibility of conducting worldwide military operations in a greatly reduced time frame.

The evolution of space operations is comparable to the development of air warfare, which similarly exploited inherent advantages in altitude and speed. Aircraft filled an essential role in supporting the ground and naval forces in the First World War through observation, antiobservation, ground attack, and communications. Between the wars, larger aircraft came into service in the form of civil and military transport, a capability that was greatly expanded during World War II.

Moreover, between the world wars the United States and Great Britain developed airpower as a means of leapfrogging conventional ground and naval battles to enable direct strikes on the enemy's ability to wage war. Although this theory of strategic bombing met with limited success in the Second World War, the concept culminated in the development of the intercontinental nuclear deterrent after the Second World War.

Space operations, like air operations in the First World War, currently provide support essential for the successful operations of terrestrial forces. Satellites enable near-real-time, worldwide communications, sensing, timing, and navigation. These capabilities, analogous to the roles of the observation balloons and aircraft of the First World War, may make possible dominant battlefield awareness and coordination of a global precision strike architecture.

An effective antisatellite (ASAT) capability could lead to the ability to achieve aerospace control or superiority in order to deny an opponent the ability to operate in or from space. An ASAT system would follow in the footsteps of the first fighter aircraft that dueled for control of the air over the trenches of the First World War and is a logical extension of the current role of air superiority fighters and developing theater air defense systems.

However, space operations will also greatly differ from air operations. First, the "geography" of space is fundamentally different from that of the earth's atmosphere. Orbital mechanics require operating speeds (17,000 miles per hour) that far surpass those currently achievable in the atmosphere.

Thus, if properly placed and employed, space assets could perform missions in much less time than state-of-the-art aircraft. One possible mission is to use space forces to project power to directly achieve national objectives (operational or strategic) in a particular theater. Space strike systems based on satellites or on transatmospheric vehicles could enable precision strikes whose quantitative advantage in speed would result in a qualitative difference in capability.

Although currently limited, future capabilities in space transport may also make possible the movement of critical forces and equipment from CONUS to a theater in time frames an order of magnitude faster than with current sea and air transport. Thus, space operations may provide important advantages in time-critical situations.

Further, the altitude advantages provided by space greatly improve surveillance and reconnaissance coverage of the earth and, as a result, could offer the means to command and control operations in theaters where distance and terrain complicate or confound terrestrially based systems.

Space, however, does have limiting factors that could constrain its military use. First, space is not amenable to human life, thus limiting the manned presence in future space operations. As a result, most of the improvements in future space operations will most likely come through unmanned technologies. In addition, the speeds associated with space flight and the amounts of fuel required to maneuver in orbit using current technologies and energy sources greatly limit the flexibility of spacecraft in orbit.

Therefore, sizable technical hurdles have to be overcome before space-based strike, antisatellite systems, spacelift, and space transport become militarily usable capabilities. Systems that could enable future space operations might include trans-atmospheric vehicles (TAVs), single-stage-to-orbit (SSTO) launch vehicles, space-based directed-energy weapons (DEW) or kinetic energy weapons (KEW), space-based ballistic missile defense (BMD), satellite defense systems, small satellites, and both space-based and ground-based distributed networks to reduce the vulnerability of space capabilities.

However, new technologies such as new materials to reduce weight, more heat-resistant and stress-resistant materials, and sources of energy more powerful and efficient than chemical reactions may be needed to make these systems truly effective.

Space warfare will likely become its own warfare area only when there is need to conduct military operations in space to obtain solely space-related goals (not missions that are conducted to support earth-based operations). For example, if the United States becomes dependent on resources unique to space (such as He_3 on the moon), it may be forced to develop technologies and operational concepts to support/defend space-based industries, command and control nodes, or colonies that are entirely non-earth dependent. In such situations, space operations would be altogether removed from any congruence with traditional air operations and would undoubtedly become a distinct warfare area.

Implications: Dominant Battlespace Awareness

Though our understanding of the unfolding RMA and its potentially new warfare areas is still evolving, several possible implications have begun to come into focus. One of our working hypotheses is that the truly revolutionary effects will come from the combination of two or more new warfare areas.

For example, some combination of space and information warfare may provide certain advantages heretofore impossible to generate. As a result, the United States may be able to achieve a degree of information dominance over an enemy by both significantly degrading his information flow and enhancing ours, thereby gaining a potential step-function increase in our information capabilities.

The application of such a system of systems may generate significantly greater capabilities in times of conflict. One potential result may be the ability to generate dominant battlespace awareness (DBA) over a particular enemy, in a particular conflict. This awareness would not magically provide perfect intelligence, but would allow the United States to detect all observable phenomenology while limiting the enemy's knowledge.

This information would translate into the ability to know force locations and characteristics (including distinguishing between targets and decoys and among target types) at all times. Furthermore, the DBA architecture could include mechanisms for disseminating this data directly to the appropriate strike systems and conducting constant battle damage assessments (BDA).

An advantage such as DBA would probably require a large percentage of the total US sensing, analysis, and data transmission assets. As a result, to generate dominant battlespace awareness for a given conflict would require borrowing from both national assets and those assets dedicated to other theaters.

In the collection arena, the constant monitoring required may increasingly emphasize airborne and terrestrial sensors rather than space-based platforms. We may well see stealthy unmanned aerial vehicles (UAVs), flying at very high altitudes, conducting a greater share of the collection duties than do manned aircraft.

Advancements in information processing should enable faster analysis of the data, while dissemination can be directly linked to the shooters. Efficiency in military strike operations should be enhanced by such an across-the-board improvement in our reconnaissance-strike architecture.

Having established the possibility of generating DBA, what might its implications be in a future conflict with a regional competitor? The enemy's goal might be to achieve a breakthrough quickly and push the US forces out of the country before they could be reinforced. This would be especially true if the enemy knew of our ability to generate DBA and feared the consequences. The United States would presumably desire to stop the attack as soon as possible, with as few losses as possible. Roughly 24 hours of warning before the launch of a standing-start attack would improve the prospect of the United States generating battlespace awareness.

A capability such as DBA could affect both the systems the United States fields and the operational concepts designed to employ them in such a conflict. One likely implication may be that a force with a greater number of long-range strike systems, tied to DBA, would be far more lethal in attriting enemy forces than would traditional forces. If the value of DBA can be shared throughout the force, then the entire time line of the conflict from locating targets, determining the best time to strike them (e.g., when they are on the move), striking them, and then assessing the success of the attack could eventually become seamless. Thus, the efficiency of US precision strike campaigns could increase substantially as current problems such as prompt targeting, selection of the proper munitions, reallocation of assets, and near-real-time BDA begin to dissipate.

On the other hand, a force designed to maximize the impact of DBA might exacerbate some difficulties faced by today's forces. First, the volume of targets made available through DBA could simply overwhelm US strike capabilities. A force heavily weighted toward long-range precision-strike weapons may not completely overcome this problem but may still provide an order-of-magnitude increase in the force's lethality. Second, such a fire-intensive force will require a very large inventory of munitions. Both of these problems, last seen in the Gulf War, may not go away.

As noted above, dominant battlespace awareness may also provide benefits that extend beyond simply increasing the effectiveness of long-range strikes. For example, another force structure implication of DBA may be the ability to truly do more with less. If DBA can tell us where the main axis of attack is coming, we may be able to use smaller forces to blunt the attack because they will be covered more effectively by fire support, and the commander will have the ability to commit reserves precisely where and when they are needed. This could lead to much improved loss/exchange ratios and the opportunity to direct certain assets against other high-value targets much earlier in a conflict.

Similarly, far lighter forces could also be used for dominating maneuvers, either airmobile or amphibious, with the goal of dislodging the enemy and allowing DBA-cued strikes to target them more easily on the move. These maneuvers can also be far less risky because the planners can select landing objectives they know to be free of enemy forces. Finally, a smaller force may be capable of exploiting a successful defense with a counteroffensive far sooner than today because we could identify the path of least resistance and focus our fire support and commit our reserves more precisely and in a more timely manner.

At the same time, we would have to recognize that there would be several categories of targets where DBA would not have a large impact. One example may be a country with large inventories of nuclear, biological, or chemical (NBC) weapons. Use of these weapons may take the form of limited chemical attacks on ports and airbases or could even include nuclear attacks against US forces. Almost certainly, DBA will not help the United States gauge the enemy's intentions vis-à-vis NBC use. Furthermore, the deep underground targets that typically house NBC systems and infrastructure would be impossible for DBA to penetrate.

Another potential area in which DBA's impact may be limited is close battle. Even after a highly successful attrition campaign, US forces may inevitably run into residual enemy forces on the ground and the resulting battle may be too confined for DBA-cued fires to be of much utility. Nevertheless, if the above implications are borne out, DBA could allow the United States to defeat an enemy quicker and with fewer losses than is currently possible.

Conclusions

This description of the revolution in military affairs is neither definitive nor conclusive. The discussion is intended primarily to stimulate thinking—thinking in unique and more meaningful ways about how warfare in the twenty-first century may be fundamentally different than it is today and, of equal importance, evaluating what we should be doing now to prepare ourselves for that eventuality.

We expect that the true revolutionary impact of future changes in the conduct of warfare will come from the intersection of precision strike, information warfare, dominating maneuver, and space warfare. Military operations in all four warfare areas will be integrated into an overall operational plan that will be decisive in terms of the course—if not the outcome—of the war.

Precision strike will hold an enemy at a distance and blind and immobilize him by destroying operationally and strategically crucial, time-urgent targets. Information warfare will deny an enemy critical knowledge of his own—as well as our—forces and turn his "fog of war" into a wall of ignorance.

Dominating maneuver will deploy the right forces at the right time and place to cause the enemy's psychological collapse and complete capitulation. Space warfare will enable the United States to project force at dramatically increased speeds in response to contingencies while denying the enemy the ability to do the same. At least that is the overall concept. What we need to do now is develop the details of how we can conduct such warfare against various categories of competitors.

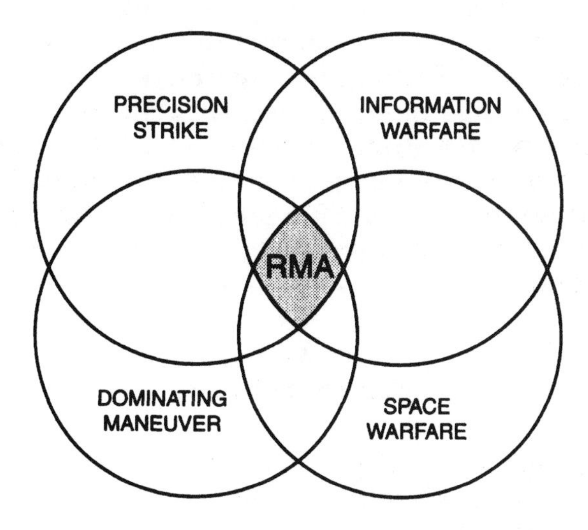

Figure 1. Elements of the Present RMA

A number of changes must occur if the United States military is going to compete successfully on the battlefields of 2020. First, there must be a change in outlook—a change in the way we think about preparing for the future. The military must nurture an attitude that supports free thinking, that accepts honest mistakes, that encourages experimentation, that rewards risk takers, and that makes provisions for starting over. As an organization, the military must break out of the box, must consider alternative futures, must think the unthinkable, and must let go of the conventional modes of operation.

Why are these changes so important? What may be the consequences if we fail to change? First, failure to change will ensure that we will not gain the most that we possibly could from the unfolding RMA. Failure to change will make it difficult for our military to make the best possible use of the new emerging technologies.

Failure to think imaginatively about the future may result in a failure to maintain the military advantages our forces so clearly demonstrated during the Gulf War. Approaches that are not innovative may prove adequate in the short term, but in the long run they may squander potential advantages needed against future competitors. We must alter our thinking and our approaches to planning if we are to be prepared for either the emergence of a large peer RMA competitor or the surprise of a true niche competitor.

To be successful and lasting, the change must come from the top—from leadership. The impetus for change must flow through the entire organization, especially through the education system. The required changes cannot occur without the support and encouragement of leadership and the enthusiasm and cooperation of the entire organization.

To be successful, we need to develop a broad strategy—a strategy robust enough to encompass and cope with the massive uncertainty we face. We need to be clear (if not always explicit) about our goals, vis-à-vis our allies as well as our competitors. We need to think through what we would like the future to look like and develop a strategy for shaping it to that end.

None of this will be easy. But a concerted, sustained, and focused effort by the Defense Department could pay dividends in the decades to come in ways as yet unforeseen.

Notes

1. Paul Bracken, "The Military After Next," *Washington Quarterly*, 16, no. 4 (Autumn 1993) 157–174.

2. Stephen Peter Rosen, Briefing on Future Competitors, US Army Roundtable Conference on the Revolution in Military Affairs, HQ US Army TRADOC, Fort Monroe, Virginia, 27 September 1993.

3. Department of Defense Directory of Military and Associated Terms, 1 December 1989, 218. See also, JCS Pub 1-02.

Overview: Future Airpower and Strategy Issues

Strategy is the art and science of translating national security objectives into practical military plans and operations. Strategy formulation in an age of revolution in affairs is especially challenging. Our two strategy essays by two premier military thinkers, Col John Warden and Col Richard Szafranski, debate the issue of just how innovative strategy formulation must be in the midst of a military revolution. Col Warden argues that "war in the twenty-first century will be significantly different for the United States from anything encountered before the Gulf War." However, Col Szafranski contends that "there may not be really much that is revolutionary in contemporary notions of parallel war and hyperwar."

Col Warden believes that twenty-first century strategy will have to ensure great precision: high casualties will not be politically tolerable; collateral damage must be minimized; nonlethal weapons will have wide application; and manipulation of information will be critical. Strategy will have to concentrate on an enemy's entire system of organization and activity, not simply its armed forces.

Using the five-ring analogy, Col Warden proposes that strategy should target an adversary's leadership, energy or resources, infrastructure, population, and armed forces. This would make airpower the dominant instrument of such new era warfare. Simultaneously attacking these essential components rather than concentrating solely on enemy armed forces is the essence of Warden's strategy. Warden believes that the five-ring analysis gives us a good picture of what to strike, and that we must view "the enemy as a system, not an independent mass of tanks, aircraft, or dope pushers." Warden's goal is "to make the cost—political, economic, and military—to the enemy higher than he is willing to pay, or to impose strategic or operational paralysis on him so that he would become incapable of acting."

Col Szafranski, however, questions whether proposals like Col Warden's aiming at the "simultaneous reduction of the enemy systems overall energy level, so that the organic system goes into shock" are really new. Is attacking the various centers of gravity in "parallel" and with "hyper" speed really a *new theory* of war? Szafranski argues that "simultaneous and integrated attacks have long been the goal of combined arms. Attacks on the leader and leadership are not new goals of warfare, whether the enemy was viewed as a system, or not, in the past." Col Szafranski adds that the nuclear attack single integrated operations plan (SIOP), long used by the US Air Force Strategic Air Command, "promoted and planned for parallel war and hyperwar long before 'the five rings' came into vogue."

While Col Warden's essay implies that the five-rings strategy can work against very large states like China, or versus almost any adversary, Col Szafranski limits its utility to smaller industrialized states and is dubious regarding utility against terrorist or insurgent organizations. "Worse," writes Szafranski, "airpower cannot make the decisive and dominant contribution to [counterterrorism and counter-insurgency] much to the chagrin of airpower advocates."

Col Warden was a key planner and organizer of the allied air campaign that gave the Coalition air superiority over Saddam Hussein's air force in the 1991 Gulf War. Parallel war and hyperwar did shock and paralyze the Iraqi state and its military forces. Warden's essay suggests that this same parallel war and hyperwar approach, emphasizing airpower as the key to rapid and complete victory, can and should be applied in future conflicts in the twenty-first century.

However, Col Szafranski argues that this theory of fighting wars, and the central role assigned to airpower in it, does not apply to all kinds of conflicts. Szafranski contends that we still have not found a theory of airpower and air strategy that applies universally. Indeed, this leads to the question of whether or not *one* air doctrine and military strategy for all contingencies can be effective across the entire spectrum of types of conflicts and types of adversaries that the United States and its allies may confront in the future. Perhaps, instead, there should be a search for multiple air doctrines for various alternative types of conflicts and enemies.

The reader can decide for himself or herself what is new or different in Col Warden's "air theory for the twenty-first century," and whether the same strategy will work across the spectrum of conflicts. What both Col Warden and Col Szafranski *do* agree on is that "paralyzing" or neutralizing an enemy's critical warfighting assets—whether military or civilian, whether physical or psychological—is important for strategy success in twenty-first century warfare.

Chapter 4

Air Theory for the Twenty-first Century

Col John A. Warden III, USAF

War in the twenty-first century will be significantly different for the United States from anything encountered before the Gulf War. American wars will be increasingly precise; imprecision will be too expensive physically and politically to condone. Our political leaders and our citizenry will insist that we hit only what we are shooting at and that we shoot the right thing. Increased use of precision weapons will mean far less dependence on the multitudes of people or machines needed in the past to make up for inaccuracy in weapons. Precision will come to suggest not only that a weapon strike exactly where it is aimed, but also that the weapons be precise in destroying or affecting only what is supposed to be affected. Standoff and indirect-fire precision weapons will become available to many others, which will make massing of large numbers in the open suicidal and the safety of deploying sea-based or land-based aircraft close to a combat area problematic.

We might hope that more accurate weapons would drive potential enemy leaders to be less enamored of achieving their political objectives with force; if we are very lucky, perhaps the world will move in this direction. Of at least equal likelihood, however, states and other entities will turn to other forms of warfare—such as attacks on enemy strategic centers of gravity. These attacks may be via missiles, space, or unconventional means, but all will recognize that they must achieve their objective before the United States chooses to involve itself. This, in turn, will increase the premium on American ability to move within hours to any point on the globe without reliance on en route bases.

The advent of nonlethal weapons technology will expand our options over the full spectrum of war. These new weapons will find application against communications, artillery, bridges, and internal combustion engines, to name but a few potential targets. And of greatest interest, they will accomplish their ends without dependence on big explosions that destroy more property than necessary and that cause unplanned human casualties. Can these weapons replace traditional lethal tools? In theory they can, as long as we accept the idea that war is fought to make the enemy do your will. What we will surely find, however, is that these weapons give us operational concepts and opportunities well beyond what would be possible if we merely substitute them for conventional weapons.

The United States can achieve virtually all military objectives without recourse to weapons of mass destruction. Conversely, other states, unable to afford the hyperwar arsenal now the exclusive property of the United States, will at least experiment with them. The challenge for America is to decide if it wants to negate these weapons without replying or preempting in kind. Accompanying this question is the question of nuclear deterrence in a significantly changed world. Although deterrence will certainly be greatly different from our cold war conception of it, does it lose its utility in all situations? How should US nuclear forces be maintained? This entire matter deserves serious thought, soon.

Information will become a prominent, if not predominant, part of war to the extent that whole wars may well revolve around seizing or manipulating the enemy's datasphere.[1] Furthermore, it may be important in some instances to furnish the enemy with accurate information. This concept is discussed later in this chapter.

The world is currently experiencing what may be the most revolutionary period in all of human existence with major revolutions taking place simultaneously in geopolitics, production, technology, and military affairs. The pace of change is accelerating and shows no sign of letting up. If we are to succeed in protecting our interests in this environment, we must spend more time than ever in our past thinking about war and developing new employment concepts. Attrition warfare belongs to another age, and the days when wars could be won by sheer bravery and perseverance are gone. Victory will go to those who think through the problem and capitalize on every tool available—regardless of its source. Let us begin laying the intellectual framework for future air operations.

All military operations, including air operations, should be consonant with the prevailing political and physical environment. In World War II the United States and her Allies imposed widespread destruction and civilian casualties on Japan and Germany; prior to the Gulf War, a new political climate meant that a proposal to impose similar damage on Iraq would have met overwhelming opposition from American and coalition political leaders. As late as the Vietnam War the general inaccuracy of weapons required large numbers of men to expose themselves to hostile fire in order to launch enough weapons to have some effect on the enemy; now, the new physical reality of accurate weapons means that few men need be or should be exposed.

Military operations must be conducted so as to give reasonable probability of accomplishing desired political goals at an acceptable price. Indeed, before one can develop or adopt a concept of operations, an understanding of war and political objectives is imperative.

For war to make any sense, it must be conducted for some reason. The reason may not be very good or seem to make much sense, but with remarkably few exceptions, most rulers who have gone to war have done so with the objective of achieving something—perhaps additional territory, a halt to offensive enemy operations, avenging an insult, or forcing a religious conversion. Very few have gone to war to amuse themselves with no concern for the outcome or desire for anything other than the opportunity to have a good donnybrook.

This is not to say, however, that all those who have gone to war have done so with a clear idea of their objective and what it would take to achieve it. Indeed, failure to define ends and means clearly has led to innumerable disasters for attacker and attacked alike. First rule: if you are going to war, know why you are going. Corollary to the first rule: have some understanding of what your enemy wants out of the war and the price each of you is willing to pay. Remember: war is not quintessentially about fighting and killing; rather, it is about getting something that the opponent is not inclined to hand over. Still another way to express this idea is this: war is all about making your enemy do something you want him to do when he doesn't want to do it—and then preventing him from taking an alternative approach which you would also find unacceptable.

There are a variety of ways to make an enemy do what you want him to do. In simple terms, however, there are but three: make it too expensive for the enemy to resist, with "expensive" understood in political, economic, and military terms; physically prevent an enemy from doing something by imposing strategic or operational paralysis on him; or destroy him absolutely.

The last of these options is rare in history, difficult to execute, fraught with moral concerns, and normally not very useful because of all the unintended consequences it engenders. We will pass over it and concentrate on the first two.

When we talk about making something so costly for an enemy that he decides to accept our position, we are talking about something very difficult to define or predict precisely. After all, human organizations typically react in an infinite number of ways to similar stimuli. The difficulty of defining or predicting, however, does not suggest that it is a hopeless task. Imprecise, yes; hopeless, no.

We all know from our experience that we regularly make decisions whether or not to do something. We don't go on a trip if it costs more money than we are ready to pay; we don't go mountain climbing if we fear the cost of falling; and we don't drive above the speed limit if the probability of a ticket seems high, and so on. Enemies, whether they be states, criminal organizations, or individuals, all do the same thing; they almost always act or don't act based on some kind of cost-benefit ratio. The enemy may not assess a situation the way we do, and we may disagree with his assessment, but assessments are part and parcel of every decision. From an airpower standpoint, it is our job to determine what price (negative or positive) it will take to induce an enemy to accept our conditions. To do so, however, we need to understand how our enemies are organized. One might object that understanding how our enemies are organized is an impossible task, especially if we don't know in advance who they are. Fortunately, this is not the case; as we shall see, every life-based system is organized about the same way. Only the details vary.

Whether we are talking about an industrialized state, a drug cartel, or an electric company, every organization follows the same organizational scheme. This is very important to us as military planners because it allows us to develop general concepts not dependent on a specific enemy. Likewise, as we understand how our enemies are organized, we can easily move on to the concept of centers of gravity. Understanding centers of gravity then allows us to make reasonable guesses as to how to create costs which *may* lead the enemy to accept our demands. If the enemy does not respond to imposed costs, then this same understanding of organization and centers of gravity shows us how to impose operational or strategic paralysis on our enemy so he becomes incapable of opposing us. Let's start with the basics of organization (table 1).

Table 1

SYSTEM ATTRIBUTES

	Body	State	Drug Cartel	Electric Company
Leader	Brain -eyes -nerves	Government -communication -security	Leader -communication -security	Central Control
Organic Essential	Food/oxygen -conversion via vital organs	Energy (electricity, oil, food), money	Coca source plus conversion	Input (heat, hydro) Output (electricity)
Infrastructure	Vessels, bones, muscles	Roads, airfields, factories	Roads, airways, sea lanes	Transmission lines
Population	Cells	People	Growers, distributors, processors	Workers
Fighting Mechanism	Leukocytes	Military, police, firemen	Street soldiers	Repairmen

As can be seen from the preceding table, a wide variety of systems ranging from an individual to an electric company are organized with remarkable similarity. This organizing scheme is sufficiently widespread to make it an acceptable starting place for working out most military or business problems. It helps us put into effect injunctions from ancient Greek and

Chinese alike to "know thyself" and "thine enemy." In addition to simplifying the "knowing" process, this organizational scheme gives us an easy way to categorize information, which we must do if we are to make real decisions. For practical purposes, the world contains an infinite amount of information which by definition cannot be totally correlated. Filters of some sort are a necessity; this systems approach provides an easy way to categorize information and to understand the relative importance of any particular bit.

Our primary interest is not in building a theory of organiza-ion; rather, it is to derive an understanding of what we might need to impose an intolerable cost or strategic or operational paralysis on an enemy. To grasp the essence of this problem, it helps to rearrange our table in the form of five rings (fig. 2).

Rearranging the tabular information into the five rings diagram gives us several key insights. First, it shows us that we are dealing with an interdependent system. That is, each ring has a relationship with all of the others and all play some role. Seeing the enemy as a system gives us enormous advantages over those who see him merely as an army or air force, or worse yet, as some quantity of tanks or airplanes or ships or drug pushers without ever understanding what it is that allows these tanks or ships to operate and for what purpose.

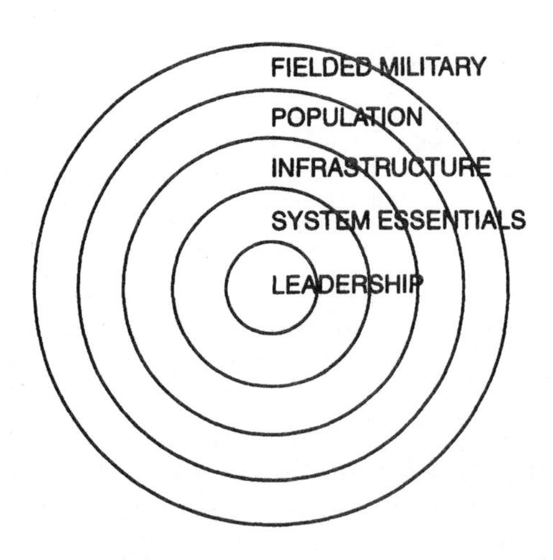

Figure 2. System Model.

Second, it gives us some idea of the relative importance of each entity contained within a given ring. For example, the head of a drug cartel (the leadership ring) has the power to change the cartel considerably whereas the street soldier (in the fielded military forces ring) assigned the job of protecting a pusher in a back alley can have virtually no effect on the cartel as a whole.

Third, it portrays rather graphically an ancient truth about war: our objective is always to convince the enemy to do what we want him to do. The person or entity with the power to agree to change is the leader in the middle. Thus, directly or indirectly, all of our energies in war should be focused on changing the mind of the leadership.

Fourth, our rings clearly show that the military is a shield or spear for the whole system, not the essence of the system. Given a choice, even in something so simple as personal combat, we certainly wouldn't make destruction of our enemy's shield our end game. *Contrary to Clausewitz, destruction of the enemy military is not the essence of war; the essence of war is convincing the enemy to accept your position, and fighting his military forces is at best a means to an end and at worst a total waste of time and energy.*

Fifth, and last, the rings give us the concept of working from the inside to the outside as opposed to the converse. Understanding this concept is essential to taking a strategic rather than a tactical approach to winning wars.

In using the rings to develop war ideas, it is imperative to start with the largest identifiable system. That is, if the immediate problem is reversing the effects of an invasion, one would start the analysis with the largest possible look at the system description of the invading country. An example: when the Iraqis invaded Kuwait, Gen Norman Schwarzkopf quickly grasped the idea that his problem was first with Iraq as a state and only secondarily with Iraq's military forces within Kuwait itself. At some point, however, we wanted to understand details about Iraq's army in Kuwait. Not surprisingly, we found that it was organized on the five-ring principle and insofar as our objective with respect to that army was something other than pure destruction, five-ring analysis gave us a good picture of what to strike. Had we so desired, we could have continued our analysis down to the level of an individual soldier because he is organized about the same way as is his country. From a diagraming standpoint, then, we start out with the big picture of the strategic entity.[2]

When we want more information, we pull out subsystems like electrical power under system essentials and show it as a five-ring system. We may have to make several more five-ring models to show successively lower electrical subsystems. We continue the process until we have sufficient understanding and information to act. Note that with this approach, we have little need for the infinite amount of information theoretically available on a strategic entity like a state. Instead, we can identify very quickly what we don't know and concentrate our information search on relevant data.

For the mathematically inclined, it will be clear that we are describing a process of differentiation as opposed to integration. In a complex world, a top-down, differentiation approach is a necessity. Important to note, however, is that virtually all our military training (and business training) starts us at the lowest possible level and asks us to work our way up. Thus, we learn a tactical approach to the world. However, when we want to think not about fighting wars but about winning them, we must take a strategic and operational—or top-down—approach if we are to succeed.

So far, we have not talked explicitly about centers of gravity, but we have derived them by showing how we and our enemies are organized. Centers of gravity are primarily organizational concepts. Which ones are most important becomes clear when we decide what effect we want to produce on the enemy in order to induce him to accept our position. Which ones to attack becomes a matter of our capability.

Let us review key concepts discussed to this point. First, the object of war is to induce the enemy to do your bidding. Second, it is the leadership of the enemy that decides to accommodate you. Third, engagement of the enemy military may be a means to an end, but the engagement is never an end in itself and should be avoided under most circumstances. Fourth, every life-form-based system is organized similarly: a leadership function to direct it, a system-essential function to convert energy from one form to another, an infrastructure to tie it all together, a population to make it function, and a defense system to protect it from attack. Fifth, the enemy is a system, not an independent mass of tanks, aircraft, or dope pushers. And sixth, the five rings provide a good method for categorizing information and identifying centers of gravity.

We said earlier that our goal in war would generally be to make the cost—political, economic, and military—to the enemy higher than he was willing to pay or to impose strategic or operational paralysis on him so that he would become incapable of acting. Now, with an understanding of how enemies are organized, we can begin the process of determining how to accomplish either or both with airpower tools.

The object of war is to convince the enemy leadership to do what you want it to do. The enemy leadership acts on some cost/risk basis, but we can't know precisely what it might be. We can, however, make some reasonable guesses based on system and organization theory. To do this, put yourself in the center of the five rings as the leader of a strategic entity like a drug cartel or state. You have certain rather basic goals that normally will take precedence over others. First, you want to survive personally (this is not to say you won't die for your system, but you probably see yourself and the system as being closely tied together). For you to survive personally (in most instances) the system you lead must survive in something reasonably close to its present form.

Let us say that you are the leader of a drug cartel and an enemy threatens you credibly with the following (to which you cannot respond): your bank accounts will be zeroed, your communications with the world outside your mountain retreat will be severed, your cocaine processing facility will be destroyed, and your house will be converted to rubble. To avoid these nasty things, all you have to do is agree to stop selling cocaine in one country. What do you do? If you are remotely rational, you agree immediately. Failure to do so means your system effectively ceases to exist, which leaves you personally in a precarious position and unable even to retire in splendor because you can't get at the billions you had socked away in a country with strict privacy in banking laws.

Suppose that only some of these dire events were threatened or deemed likely. In this case, you might choose to negotiate. Perhaps you would agree to sell less cocaine in the target country. Perhaps you would agree to a moratorium on sales. Your enemy might or might not accept these counterproposals; it would depend largely on how much he was willing to spend to create a cost for you. In our ideal world we like to think we don't negotiate with drug dealers or tyrants like Saddam Hussein. In the real world we do so all the time. *Very rarely are we willing to invest the time and effort required to achieve maximal results.*

Our discussion of costs has so far been oriented at a strategic level. Does it also apply at an operational level—the level at which military forces are actually employed? The answer is an absolute yes. Military commanders, with the exception of a few really stupid ones, have always weighed costs as they were planning or conducting an operation. Let's take a hypothetical look at George Patton and the Third Army in World War II.

George Patton was an aggressive commander who believed that speed of advance was key to success. Obviously, then, the Third Army needed to move quickly as a system—not just the tanks, but the whole system that supported them at the front. From the German side, if moving fast was good from George Patton's perspective, it was bad from theirs. Now, let us do a quick five-ring analysis of the Third Army from just the cost standpoint. (We will return to it later when we discuss operational paralysis.)

Let us suppose that something catastrophic happened to Third Army's fuel supply in mid-September of 1944. Let us assume that someone tells General Patton at a staff meeting that all fuel deliveries to his Army will cease in two days. His choices are basically two: slow down or stop the movement of his army so that it can assume a reasonable defensive position, or tell everyone to plunge ahead as far as possible until they run out of gas. Since the latter is likely to leave the majority of the army in an untenable and unplanned position and is unlikely to achieve anything final, Patton opts for the former because he has assessed the cost of continuing as too high for the possible results.

Realize also that unbeknownst to the commanding general, every subordinate commander and soldier will start acting on information about an impending fuel shortage as soon as he hears about it. The effect is obvious; by the time the formal order to halt comes down, forward movement already would have ceased and hoarding of remaining fuel supplies would have become widespread. The principle is simple: at all levels, leaders make decisions based on a cost/benefit analysis.

Before moving on to discuss imposition of strategic and operational paralysis, we need to make two more points on the subject of cost. The first is that the enemy leader may not recognize how much attacks on him are costing at the time of attack and in the future. This almost certainly was the case with Saddam Hussein, who simply failed to comprehend for several weeks what the strategic air attacks against him were doing to his future. Had he understood, he might have sued for peace the first morning of the war. His lack of understanding flowed from ignorance of the effect of modern air attack[3] and from lack of information. The coalition attack in the first minutes had so disrupted communications at strategic levels that it was very difficult to receive and process damage reports.[4]

A similar event may have taken place in Japan in late 1944 and into 1945. Japanese army leaders persisted in their desire to continue the war even though their homeland was collapsing around them as a result of strategic air and sea attacks. They apparently lacked the in-depth understanding of war and their country to appreciate what was really happening. Like Saddam a half-century later, the Japanese were stuck in a paradigm that said that the only important operations in war were the clashes of armies. In the Japanese case part of the problem may have stemmed from the Bushido code of personal bravery that tended to assume that success in war would be a function of agglomerating many tactical successes. The concepts of strategic and operational war were simply not there.

Two lessons flow from these examples: you may have to educate the enemy on the effect your operations are likely to have. You may also have to give him accurate information on the extent of his losses—*and the long-term and short-term effects likely to flow from them.*

As we have seen, we cannot depend on the his making the concessions we ask because of a realistic cost/benefit analysis. In the event we cannot educate and inform him properly, we must be ready to consider imposition of strategic and operational paralysis. Fortunately, the effort we put into understanding how enemies are organized and how to impose costs leads us directly into the concepts and mechanisms of paralysis.

The idea of paralysis is quite simple. If the enemy is seen as a system, we need to identify those parts of the system which we can affect in such a way as to prevent the system from doing something we don't want it to do. The best place to start is normally at the center for if we can prevent the system's leadership from gathering, processing, and using information we don't want him to have, we have effectively paralyzed the system at a strategic level.

Let us go back to the drug cartel example we earlier discussed. Suppose that the suppliers and pushers hear nothing from headquarters for some period of time. Their finances begin to dry up, nobody protects them from competitors, and their stocks dwindle quickly. What do they do? They begin to look for other cartels to deal with or for other lines of work. In a short period of time, but not instantly, the paralysis imposed at the strategic level of the cartel destroys the organization's ability to sell the drugs its opposition didn't want it to sell—all while the overwhelming majority of the individuals in the organization are unharmed and not even directly threatened.

The obvious place to induce strategic system paralysis is at the leadership, or brain, level. What happens, however, if the brain cannot be located or attacked? Although the leadership function always provides the most lucrative place to induce paralysis, it is not the only possibility. Suppose that we can't reach the drug lord, but we can reach and destroy one of his system essentials, such as his financial net? We are likely to have created a different level of paralysis. The organization may still be able to function, and it will certainly search furiously to repair or replace its financial net. If it doesn't succeed, however, this paralysis in one part of the strategic system is likely to cause much of the rest to atrophy and become ineffective. After all, the majority of the organization's suppliers and workers must eat and must pay for services rendered. If they are not getting regular pay, they are going to be forced again to find alternatives outside the organization that can no longer provide them with a system essential.

At a big-state level, one can imagine a similar outcome if the state loses a system essential like electricity. Imagine the effect on the United States if all of its electricity stops functioning.

Let us return to our George Patton and Third Army example to look at operational paralysis. Patton depended on speed for success, but not unfocused speed. He needed to know where he was going, what his troops were going to encounter, where to send the fuel and ammunition, and where to shift land and air forces as required. Suppose the Germans had succeeded in blinding Patton by depriving him of his ability to gather and disseminate information. Under these conditions, Third Army would have been effectively paralyzed insofar as its ability to conduct rapid offensives simply because offensives on the ground at any speed are extraordinarily complex and require huge amounts of good information.

If the Germans were unable to blind Patton, what else could they have done to induce operational paralysis? Again, going from the center of the rings toward the outside suggests we should look next in the system-essential (or supply) ring for an answer. Most assuredly, Patton's speed depended on fuel for his tanks and trucks—no fuel, no speed. Thus, elimination of fuel, perhaps by interdicting the Red Ball Express, induces the desired form of operational paralysis and converts Third Army into something different. Before its fuel was cut off, Third Army was a fast moving, dangerous threat to the Germans; after the fuel stops, it becomes a slow, slogging beast significantly different in nature and threat.

So far we have discussed effects we might want to produce on the enemy: untenable costs or paralysis at one or more levels. Next, we must look at how we go about doing it. Before proceeding, however, it is useful to note that we have used quite a few pages talking about air theory and have yet to discuss bombs, missiles, or bullets. The reason is simple; well before it makes any sense to talk about mechanics, it is imperative to decide what effect you want to produce on the enemy. Making this decision is the toughest intellectual challenge; once the desired effect is decided, figuring out how to attain it is much easier if for no other reason than we practice the necessary tactical events every day, whereas we rarely (far too rarely) think about strategic and operational problems. Let us propose a very simple rule for how to go about producing the effect: do it very fast.

It may seem facetious to reduce the "how" to such a simple rule, and, indeed, we will now make it a little more complex by talking about parallel attack. Nevertheless, the essence of success in future war will certainly be to make everything happen you want to happen in a very short period of time—instantly if possible. Why? And what is parallel war?

Parallel war brings so many parts of the enemy system under near-simultaneous attack that the system simply cannot react to defend or to repair itself. It is like the death of a thousand cuts; any individual cut is unlikely to be serious. A hundred, however, start to slow a body considerably, and a thousand are fatal because the body cannot deal with that many assaults on it. Our best example of parallel war to date is the strategic attack on Iraq in the Gulf War. Within a matter of minutes the coalition, attacked over a hundred key targets across Iraq's entire strategic depth. In an instant, important functions in all of Iraq stopped working very well. Phone service fell precipitously, lights went out, air defense centers stopped controlling subordinate units, and key leadership offices and personnel were destroyed. To put Iraq's dilemma in perspective, the coalition struck three times as many targets in Iraq in the first 24 hours as Eighth Air Force hit in Germany in all of 1943!

The bombing offensive against Germany (until the very end) was a serial operation, as virtually all military operations have been since the dawn of history. Operations have been serial because communications made concentration of men imperative, the inaccuracy of weapons meant that a great number had to be employed to have an effect, and the difficulty of movement essentially restricted operations to one or two locations. In addition, military operations have mostly been conducted against the enemy's military, not against his entire strategic or operational system. All this meant that war was a matter of action and reaction, of culminating points, of regrouping, of reforming. Essentially, war was an effort by one side to break through a defensive line with serial attacks or it was an attempt to prevent breakthrough.

In any event the majority of the enemy system lay in relative safety through most of a conflict with the fighting and damage confined largely to the front itself. Even when aerial bombardment began to reach strategic depths, the bombardment tended to be serial (again because of inaccurate weapons and the need to concentrate attacking forces so they could penetrate an aerial defensive line). This meant that the enemy could gather his defenders in one or two places and that he could concentrate the entire system's repair assets on the one or two places which may have suffered some damage. Not so in Iraq.

In Iraq, a country about the same size as prewar Germany, so many key facilities suffered so much damage so quickly that it was simply not possible to make strategically meaningful repair. Nor was it possible or very useful to concentrate defenses; successful defense of one target merely meant that one out of over a hundred didn't get hit at that particular time. As in the thousand cuts analogy, it just doesn't matter very much if some of the cuts are deflected. It is important to note that Iraq was a very tough country strategically. Iraq had spent an enormous amount of money and energy on giving itself lots of protection and redundancy and its efforts would have paid off well if it had been attacked serially as it had every right to anticipate it would. In other words, the parallel attack against Iraq was against what may well have been the country best prepared in all the world for attack. If it worked there, it will probably work elsewhere.

Executing parallel attack is a subject for another essay or even a book. Suffice it to say here that those things brought under attack must be carefully selected to achieve the desired effect.

We have now provided the groundwork for a theory of air power to use into the twenty-first century. To summarize: understand the political and technological environment; identify political objectives; determine how you want to induce the enemy to do your will (imposed cost, paralysis, or destruction); use the five-ring systems analysis to get sufficient information on the enemy to make possible identification of appropriate centers of gravity; and attack the right targets in parallel as quickly as possible. To make all this a little more understandable, it is useful to finish by mentioning the Gulf War's key strategic and operational lessons, which look as though they will be useful for the next quarter-century or more.

We can identify 10 concepts that summarize the revolution of the Gulf War and that must be taken into account as we develop new force levels and strategy:

1. The importance of strategic attack and the fragility of states at the strategic level of war
2. Fatal consequences of losing strategic air superiority
3. The overwhelming effects of parallel warfare
4. The value of precision weapons
5. The fragility of surface forces at the operational level of war
6. Fatal consequence of losing operational air superiority
7. The redefinition of mass and surprise by stealth and precision
8. The viability of "air occupation"
9. The dominance of airpower
10. The importance of information at the strategic and operational levels

Let us look at each of these briefly.

1. *The importance of strategic attack and the fragility of states at the strategic level of war.* Countries are inverted pyramids that rest precariously on their strategic innards—their leadership, communications, key production, infrastructure, and population. If a country is paralyzed strategically, it is defeated and cannot sustain its fielded forces though they be fully intact.

2. *Fatal consequences of losing strategic air superiority.* When a state loses its ability to protect itself from air attack, it is at the mercy of its enemy, and only the enemy's compassion or exhaustion can save it. The first reason for government is to protect the citizenry and its property. When a state can no longer do so, it has lost its reason for being. When a state loses strategic air superiority and has no reasonable hope of regaining it quickly, it should sue for peace as quickly as possible. From an offensive standpoint, winning strategic air superiority is the number one priority of the commander; once that is accomplished, everything else is a just a matter of time.

3. *The overwhelming effects of parallel warfare.* Strategic organizations, including states, have a small number of vital targets at the strategic level—in the neighborhood of a few hundred with an average of perhaps 10 aimpoints per vital target. These targets tend to be small, very expensive, have few backups, and are hard to repair. If a significant percentage of them are struck in parallel, the damage becomes insuperable. Contrast parallel attack with serial attack where only one or two targets come under attack in a given day (or longer). The enemy can alleviate the effects of serial attack by dispersal over time, increasing the defenses of targets that are likely to be attacked, concentrating his resources to repair damage to single targets, and conducting counteroffensives. Parallel attack deprives him of the ability to respond effectively, and the greater the percentage of targets hit in a single blow, the more nearly impossible is response.

4. *The value of precision weapons.* Precision weapons allow the economical destruction of virtually all targets—especially strategic and operational targets that are difficult to move or conceal. They change the nature of war from one of probability to one of certainty. Wars for millennia have been probability events in which each side launched huge quantities of projectiles (and men) at one another in the hope that enough of the projectiles (and men) would kill enough of the other side to induce retreat or surrender. Probability warfare was chancy at best. It was unpredictable, full of surprises, hard to quantify, and governed by accident. Precision weapons have changed all that. In the Gulf War, we knew with near certainty that a single weapon would destroy its target. War moved into the predictable.

With precision weapons, even logistics become simple; destruction of the Iraqis at the strategic, operational, and tactical levels required that about 12,000 aimpoints be hit. Thus, no longer is it necessary to move a near-infinite quantity of munitions so that some tiny percentage might hit something important. Since the Iraqi army was the largest fielded since the Chinese in the Korean War and since we know that all countries look about the same at the strategic and operational levels, we can forecast in advance how many precision weapons will be needed to defeat an enemy—assuming of course that we are confident about getting the weapons to their targets.

5. *The fragility of surface forces at the operational level of war.* Supporting significant numbers of surface forces (air, land, or sea) is a tough administrative problem even in peacetime. Success depends upon efficient distribution of information, fuel, food, and ammunition. By necessity, efficient distribution depends on an inverted pyramid of distribution. Supplies of all operational commodities must be accumulated in one or two locations, then parceled out to two or four locations, and so on until they eventually reach the user. The nodes in the system are exceptionally vulnerable to precision attack. As an example, consider what the effect would have been of a single air raid a day—even with nonprecision weapons—on the WWII Red Ball Express or on the buildup behind VII and XVIII Corps in the Gulf War. The Red Ball Express became internally unsustainable, and the VII and XVIII Corps buildups severely strained the resources of the entire US Army—even in the absence of any enemy attacks.

Logistics and administration dominate surface warfare, and neither is easy to defend. In the past these activities took place so far behind the lines that they were reasonably secure. Such is no longer the case—which brings into serious question any form of warfare that requires huge logistics and administrative buildup.

6. *Fatal consequence of losing operational air superiority.* Functioning at the operational level is difficult even without enemy interference. If the enemy attains operational air superiority (and exploits it)[5] and can roam at will above indispensable operational functions like supply, communications, and movement, success is not possible. As with the loss of strategic air superiority, loss of operational air superiority spells doom and should prompt quick measures to retreat—which is likely to be very costly—or to arrange for surrender terms.

7. *The redefinition of mass and surprise by stealth and precision.* For the first time in the history of warfare, a single entity can produce its own mass and surprise. It is this single entity that makes parallel warfare possible. Surprise has always been one of the most important factors in war—perhaps even the single most important, because it could make up for large deficiencies in numbers. Surprise was always difficult to achieve because it conflicted with the concepts of mass and

concentration. In order to have enough forces available to hurl enough projectiles to win the probability contest, a commander had to assemble and move large numbers. Of course, assembling and moving large forces in secret was quite difficult, even in the days before aerial reconnaissance, so the odds on surprising the enemy were small indeed. Stealth and precision have solved both sides of the problem; by definition, stealth achieves surprise, and precision means that a single weapon accomplishes what thousands were unlikely to accomplish in the past.

8. *The viability of "air occupation."* Countries conform to the will of their enemies when the penalty for not conforming exceeds the cost of conforming. Cost can be imposed on a state by paralyzing or destroying its strategic and operational base or by actual occupation of enemy territory. In the past, occupation (in the rare instances when it was needed or possible) was accomplished by ground forces—because there was no good substitute. Today, the concept of "air occupation" is a reality and in many cases it will suffice. The Iraqis conformed with UN demands as much as or more than the French did with German demands when occupied by millions of Germans. Ground occupation, however, is indicated when the intent is to colonize or otherwise appropriate the enemy's homeland.

9. *The dominance of airpower.* Airpower (fixed wing, helicopter, cruise missile, satellite), if not checked, will destroy an enemy's strategic and operational target bases—which are very vulnerable and very difficult to make less vulnerable. It can also destroy most tactical targets if necessary.

10. *The importance of information at the strategic and operational levels.* In the Gulf War, the coalition deprived Iraq of most of its ability to gather and use information. At the same time, the coalition managed its own information requirements acceptably, even though it was organized in the same way Frederick the Great had organized himself. Clear for the future is the requirement to redesign our organizations so they are built to exploit modern information-handling equipment. This also means flattening organizations, eliminating most middle management, pushing decision making to very low levels, and forming worldwide neural networks to capitalize on the ability of units in and out of the direct conflict area.

The information lesson from the Gulf was negative; the coalition succeeded in breaking Iraq's ability to process information but failed to fill the void by providing Iraqis an alternate source of information.[6] This failure made Saddam's job much easier and greatly reduced the chance of his overthrow. Capturing and exploiting the datasphere may well be the most important effort in many future wars.

Beyond these Gulf War lessons, which have applicability well into the future, it behooves the air planner to think of one other area: what can be done with airpower that in the past we knew could only be done with ground or sea power or couldn't be done at all? The question must be addressed for several reasons: airpower has the ability to reach a conflict area faster and cheaper than other forms of power; employment of air power typically puts far fewer people at risk than any other form (in the Gulf War, there were rarely more than a few hundred airmen in the air as opposed to the tens of thousands of soldiers and sailors in the direct combat areas); and it may provide the only way for the United States to participate at acceptable political risk (use of airpower does not require physical presence on the ground). Let us look at just one example.

Suppose a large city is under the control of roving gangs of soldiers, and it is American policy to restore some degree of order to the city. Normally, we would think that could only be done by putting our own soldiers on the ground. But what if policymakers are unwilling to accept the political and physical risks attendant to doing so? Do we do nothing, or do we look for innovative solutions?

If we define the problem as one of preventing groups of soldiers from wandering around a city, we may be able to solve it from the air. Can we not put a combination of AC-130s and helicopters in the air equipped with searchlights, loudspeakers, rubber bullets, entangling chemical nets, and other paraphernalia? When groups are spotted, they first receive a warning to disperse. If they don't they find themselves under attack by nonlethal, but unpleasant, weapons. If these don't work, lethal force is at hand. It may be very difficult to prevent an individual from skulking around a city or even robbing an occasional bank. Single individuals, however, constitute a relatively small tactical problem since they are unlikely to be able to cause wide-scale disruption as can multiple groups. The latter problem is serious but manageable; the former is a police matter.

By the same token, we know that we will be called on to conduct humanitarian and peacemaking operations. If we think about food delivery as the same as bomb delivery and understand that with food, as with bombs, our responsibility is to distribute it to the right people, we should be able to do as well with food as we do with bombs. To do so, however, will require putting as much effort into developing precision food-delivery techniques as we put into developing precision bomb or cluster-bomb capabilities. The problem is the same and is theoretically susceptible to an airpower solution if we are willing to think outside the lines. And indeed, thinking outside the lines will be a necessity if airpower is to prosper and to play a key role in defending American interests well into the next century.

Indeed, there is a new world building around us and the revolutions in politics, business, and war have happened and we must deal with them, not ignore them. Of course, it is human nature to stay with the old ways of doing business even when the external world has made the old ways obsolete or even dangerous. So many examples come so easily to mind: the heavy knights at Agincourt refusing to believe that they were being destroyed by peasants with bows; the French in World War I exulting the doctrine of "cold steel" against the machine gun and barbed wire as the flower of a generation perished; and the steel and auto makers of the United States convinced that their foreign competitors were inept even as their market positions plummeted. Accepting the changes made manifest in the Gulf War will be equally difficult for the United States but by no means impossible, if we all resolve to think.

Notes

1. A term introduced by Don Simmons in *Hyperion* (New York: Bantam Books, 1990).

2. A strategic entity is any self-contained system that has the general ability to set its own goals and the wherewithal to carry them out. A state is normally a strategic entity as is a drug cartel or a guerrilla organization.

3. Saddam made the following statement shortly after his successful invasion of Kuwait. "The United States depends on its Air Force and everyone knows that no one ever won a war from the air." Thus, his

preconceived notion (shared by many military officers around the world) made it difficult for him to analyze what was happening to him.

4. As an aside, the planners had recognized that Saddam would not be able to gather information, so they had intended to provide him accurate reports of all attacks by using psychological warfare assets. For a number of reasons, the planners were unable to make this happen; as a result, Saddam lacked information that the coalition really wanted him to have.

5. Some would argue that the mujahideen in Afghanistan lost operational air superiority and yet still prevailed. The latter is true; the former is not, because the Stinger antiaircraft missiles forced the Soviets to operate at an altitude that deprived them of the ability to hit anything. The Soviets simply did not have the precision weaponry and detection capability the United States had in the Gulf War.

6. The coalition provided Iraqi soldiers at the front great quantities of information and did so effectively; the same thing did not happen at the strategic level inside Iraq for a variety of not very good reasons.

Chapter 5

Parallel War and Hyperwar: Is Every Want a Weakness?

Col Richard Szafranski, USAF

Something happened in Desert Storm never witnessed before. Air power—in thousands and thousands of Coalition sorties—appeared to have defeated an enemy. Advocates of air power earnestly want others to accept what they believe Desert Storm proved. Desert Storm proved, they assert, that air power can be or is dominant and decisive,[1] fulfilling the vision of Gulio Douhet, Billy Mitchell, Jimmy Doolittle,[2] and others[3]—or so advocates believe and want us to believe.

That want may be a weakness. This essay examines the notions of parallel war and hyperwar, principally as they apply to air campaigns, revealing minor flaws in the ideas and their application. First, it examines parallel war to determine what is new in the idea. Second, it examines this "new" kind of air warfare to illuminate the strengths and shortcomings parallel war evidences in theory and practice. Third, it argues that the theory is useful if applied against weak industrial states. Next, it postulates theoretical ways to defeat an adversary intending to employ parallel war. Defeating parallel war is possible whether the United States is the nation intending to employ it, or whether some post-Gulf War aficionado embraces the theory. None of this intends to do any more than add greater discernment to the theory of parallel war and hyperwar. Theories of parallel warfare are not "bad" or "wrong." Rather, their shortcoming is that they have only limited utility in the emerging world.

 An edited version appears in the August 1995 issue of *Proceedings* (vol. 121, no. 8). "Is every want a weakness?" is a Zen Buddist reflection.

But, What If . . .?

Even so, one caveat remains: if parallel war is the new air warfare form, it would be a valuable one indeed. If authentic parallel war were possible, it would, as its advocates argue, render much of Clausewitz irrelevant. Clausewitz himself noted that

> if war consisted of one decisive act, or a set of simultaneous decisions, preparations would tend toward totality, for no omission could ever be rectified. . . .
>
> But, of course, if all the means available were, or could be, simultaneously employed, all wars would automatically be confined to a single decisive act or a set of simultaneous ones—the reason being that any *adverse* decision must reduce the sum of the means available, and if *all* had been committed in the first act there could really be no question of a second. Any subsequent military operation would virtually be part of the first—in other words, merely an extension of it.[4]

Three "ifs," with three "coulds" and two "woulds" in tandem, suggest skepticism. Since everything pivots on "if," the ideas bear close examination.

What Is Parallel War?

The idea of parallel war arises from understanding the enemy as a "system" or "organism," simultaneously more complicated and less complicated than the people-state-armed forces system described by Clausewitz.[5] The enemy state theoretically has five key organic components: (1) fielded military forces at the periphery; (2) the masses of the people who are not direct combatants; (3) a transportation infrastructure providing organic essentials; (4) the organic essentials themselves; and, (5) residing at the center, leadership or a controlling mechanism for the entire system. Advocates refer to these orbits or concentric "rings" as "the five rings."[6] Like a fractal, each of the rings also has within it the five components. Thus, the fielded forces at the periphery, from the army to the individual soldier, have within them leadership or some internal controlling mechanism.

In this taxonomy, the entire system and each ring has within it key nodes or "centers of gravity."[7] Leadership, the controlling mechanism, is the key node in each ring and throughout. The theory is that simultaneous and coordinated operations against all the key nodes in the system and in each of the rings are the essence of the a new kind of offensive military air campaign. Air, the theory holds, is the superior medium for prosecuting these operations. It is air power, the theorists argue, that allows attacks against the internal rings and all the other rings without first collapsing the outer rings that surround the inner ones.[8]

In contrast, serial warfare is, or was, warfare that engaged each ring and its categories of targets in turn, *ad seriatim*, moving from the periphery toward the center.[9] In the past it was not possible to attack the sovereign in the castle until the opposing army had defeated the enemy monarchs fielded forces and moved through the population toward the center. In World War II, air attempted to engage its targets in parallel, but more often engaged its air targets and target sets serially within the organic whole of Germany, giving ball-bearing factories, submarine pens, petroleum, airfields, rail and road networks, and cities some priority for air attack at any given time. Parallel war, on the other hand, theoretically employs air power to attack *all* the decisive points in each ring and *the* decisive point of the entire system *simultaneously*. The object is not just the destruction of targets. Destruction is the means to an end. The object is to destroy or damage (or to render dysfunctional) those targets that produce a strategic effect by causing loss of the enemy system's organic capabilities.[10] When these parallel warfare attacks occur with simultaneity or great speed, hyperwar results.

In a later iteration of the theory, and as information war[11] becomes an intriguing notion within the services and the Department of Defense, information becomes the "bolt" running through all the rings and holding the rings together.[12] John Warden, the leading theorist, also asserts that whatever else "weapons" may be or do, they are essentially "information" because they communicate "messages," or "meaning." Thus, air power delivers "information" to the enemy leadership. The most important information delivered in this ethereal sense is the message "stop fighting," or "your strategy has been defeated and you are paralyzed," or, in the extreme case, "you are dying."

Death or paralysis of the systems military or warfighting capability is the objective and intended effect of the air campaign. The goal is to visit the cumulative "death of a thousand cuts" on the enemy system.[13] Because the object sought is paralysis, parallel war and hyperwar aim at the sudden and simultaneous reduction of the enemy systems overall "energy level," so that the organic system goes into "shock." The simultaneous engagement of centers of gravity prevents recovery from this shock because the energy available to the system is inadequate to restore the system to full functioning. Thus, attacks against the centers of gravity must occur not only in parallel, but also with "hyper" speed. Since ground forces cannot do this, since forces afloat cannot do this (except through their air power), then air forces are the forces best suited for employing parallel war at hyperwar tempo. Or so the theory goes.

What is New Here?

According to Jeffrey R. Cooper, what might be new is a way of fighting, enabled by technology, that could evidence both coherence and simultaneity. Cooper writes:

> At the operational level, the impact of these coherent operations is to overwhelm the opponent's ability to command and control his forces, denying him the ability to respond to our campaign plan and operations, and forcing him at the limit to execute only uncoordinated preplanned actions.

The attacks themselves are

> a (massively) parallel series of synchronized integrated operations conducted at high-tempo, with high lethality and high mobility, throughout the depth and extent of the theater, intended to force the rapid collapse of both the enemy's military power and the enemy's will.

The consequence of the attacks is rapid defeat of the enemy force

> due to the simultaneous parallel operations, the high mobility, the high lethality, and the capability for sustained high tempos of operations, so many enemy units can be defeated in detail simultaneously that the operation may resemble a more classic *coup de main* executed in a single main-force engagement.[14]

But is this a new theory? Military forces since Clausewitz have been enjoined to identify and engage the "center of gravity" of an enemy's military capability. Simultaneous and integrated attacks have long been the goal of combined arms. Attacks on the leader and leadership are not new goals of warfare, whether the enemy was viewed as a system or not in the past. Even in chess, not a new game, it is possible to impose checkmate without the serial destruction of all the adversary's knights, rooks, and pawns. Nor is it novel that such a campaign theory would be advanced by airpower advocates. What the air aspects of the theory promise seem to differ little from what Douhet, Mitchell, and the faculty of the Air Corps Tactical School promised. Air power, we have always been told—even promised—by air power advocates, will be decisive. That the *United States Strategic Bombing Survey* and *The Gulf War Airpower Survey* both used empirical data to show that some of the undertakings of airpower fell short of the vision of airpower cannot be ignored.[15]

Nor should anyone ignore that the idea that parallel warfare, as distinguishable from serial warfare, is *not* a new strategic conception. As early as 1951, a naval officer, Capt (later Rear Adm) J. C. Wylie, asserted that there were two types of strategy: "sequential" and "cumulative." He described the cumulative approach in an article that appeared in *Proceedings* in 1952. Wylie later wrote that

> there are actually two very different kinds of strategies that may be used in war. One is the sequential, the series of visible, discrete steps, each dependent on the one that preceded it. The other is the cumulative, the less perceptible minute accumulation of little items piling one on top of the other until at some unknown point the mass of accumulated actions may be large enough to be critical. They are not incompatible strategies, they are not mutually exclusive. Quite the opposite. In practice they are usually interdependent in their strategic result.[16]

Moreover, the architects of the nuclear single integrated operations plan (SIOP), like Wylie, promoted and planned for instant cumulative war, or parallel war and hyperwar, decades before current theorists articulated "the five rings." As Desmond Ball has shown, SIOP nuclear weapons were allocated against target sets in the former Soviet system characterized as "leadership" (leadership), "nuclear force" (nuclear forces), "economic and industrial" (organic essentials and logistics infrastructure), and "other military" (other fielded forces).[17] The, SIOP also evidenced coherence and simultaneity, using Cooper's terms. Thus, there is scant difference between the targeting logic of the SIOP approach and the targeting logic of the five-rings approach, save for the important distinction that one employed nuclear weapons effects and the other did not, but might have.[18] While the difference between the nuclear SIOP and parallel war waged with conventional weapons is critically important, there are more similarities between the theories than differences. Both approaches sought to strike decisive points, both sought to checkmate enemy leadership, both were executed simultaneously and with hyper speed,[19] both aimed at driving down enemy "energy levels" dramatically, both sought to impose shock and paralysis on the enemy system, and both sought to eliminate rapid (or almost "any," in the case of the SIOP) enemy post-attack recovery capability.[20] Nuclear weapons use *does* make a difference. The SIOP intended to be so threatening that it also may have been self-deterring. Parallel warfare using nonnuclear appears no less threatening in terms of its immediate consequences, but has fewer constraints on its employment. Even so, the difference in weapons is not a difference in the theory *qua* theory nor in the proximate effects the SIOP and nonnuclear parallel war sought.[21]

Strengths and Shortcomings

The strength of cumulative strategies, both the SIOP and parallel war, even though they are the same theory, is that they promise to reduce more rapidly the war-making capacity of an industrialized enemy state.[22] It is indisputable that industrial states may be organized as the kind of system represented. The logic of a cumulative model appears sound, albeit somewhat mechanical, in the case of the five rings, and there are lucrative targets for air attack throughout the enemy system. The air campaign in Desert Storm demonstrated that the combat power of Iraq or a state like Iraq can be reduced by apparently simultaneous and coherent attacks against important targets. The SIOP, had it been executed, would likely also have proven the point against a more robust belligerent and war-fighting system.[23] Conventional weapons have the additional advantage of being easier to employ and having fewer constraints on their employment than nuclear weapons. Unconventional weapons have the added advantage, in some cases, of producing effects that can be reversed. Thus, the ability to prosecute these kinds of nonnuclear attacks, using SIOP targeting logic under a new name, is a valuable adjunct to warfare in the latter years of the "second wave."[24]

While a cumulative strategy promises to be effective against any enemy, one difficulty with the five-rings model is that it is ill equipped for coping with organisms that are not industrialized or industrializing state systems. Certainly a terrorist organization is a "system" that has separate component parts. Of course an insurgent organization is a "system" that has differentiated component parts. While theoretically possible to differentiate the component parts of both terrorist systems and insurgent organizations, it is not always easy actually to identify or to isolate these parts. As physical entities, the component parts, or five rings of terrorist and insurgent organization are exceedingly difficult for the air campaign planner to target. Thus, the model holds, but becomes exquisitely irrelevant for these types of organizations and counterterrorism warfare and counter-insurgency warfare. Worse, airpower cannot make the decisive and dominant contribution to these kinds of fights, much to the chagrin of airpower advocates. Fighting in the former Yugoslavia, the ill-starred intervention in Somalia, and our impotence in stopping the genocide in Rwanda are only the more recent examples of the limits of airpower. Airpower, it would seem, works best in massive doses applied as an antidote against the strength of industrial or industrializing states, uniformed armed forces, and identifiable leaders and other targets.

The five-rings model thus becomes illogical or at least impractical for nontrinitarian warfare, or what Gen John Boyd calls "irregular warfare." Nontrinitarian warfare, as described by Martin van Creveld, is warfare wherein the warring sides do not manifest the organization of Clausewitz's remarkable "trinity" of state, people, and armed forces. If, as John Keegan, Carl Builder, Martin van Creveld and others suggest, conventional war between industrial states is the less likely warfare form for the future,[25] this need not invalidate the theory entirely. The model remains extremely useful for its heuristic value to novitiate students of warfare and as a thesis to stimulate antithesis and debate.

Nontrinitarian warfare is only one of the challenges with which the theory cannot contend. Some of the characteristics of warfare on the eve of the "third wave" also confound the theory: demassification, diversity, and ninjitsu. Demassification is the fractionating of large conventional targets into much smaller ones. For example, mainframe computers are an attractive and easy-to-target set of nodes. Distributed laptops are less attractive targets because they are simultaneously more numerous, less easy to locate, mobile, and less easy to target. "Third-wave" information technology liberates leadership and leadership command centers from the requirement to reside in fixed locations. Just as telecommuting is possible for nonwarfare "knowledge workers" today, it is not inconceivable that the leaders of warfare operations in the future can command these operations from their domiciles, from nonbelligerent states, or from offshore. As hierarchies yield to networks, leadership will also become demassified. "Virtual" presence makes distance command and control possible. Thus, even among warring states, the leaders need not reside in the states to direct the fighting.

Miniaturization combines with demassification to complicate the challenge. A satellite dish receiver that measures three to five meters in diameter is an easy target for precision-guided or even area weapons. A satellite receiver or transmitter that measures one-half meter in diameter is a more difficult target to strike, especially if thousands are employed in a distributed network. While the model may be valid, the targeting challenge is such that the model might as well be invalid since it has little utility.

Dual-use technologies and facilities also confound the five-rings campaign planner. Fermentation chambers, for example, are essential components of a system that brews beer. These same fermenters are also essential for the production of biological weapons. Beer is good. Biological weapons are not good. Dual-use systems do not fit easily into the targeting template. Information technology is ubiquitous and much of it serves multiple constituencies. The Global Positioning System (GPS) would be a lucrative target, but the constellation of these satellites is demassified and distributed. Moreover, GPS users are military and civilian. Thus, the consequences of attacking GPS must be borne by friendly forces, enemy forces, and neutrals. Most communications satellites pose the same type of problem for campaign planners. Demassification, miniaturization, and dual-use also make ninjitsu—the "art of invisibility"—possible. By distributing important elements of a system, reducing them dramatically in size, embedding them in other things (religious facilities, civilian hospitals, university research centers), these elements effectively become invisible to the campaign planner.[26]

There are at least seven additional minor problems. First, the attacks may not actually occur simultaneously, except when compared to warfare of the distant past. Next, like the SIOP, the current model strives for the "decisive battle" in new form. Third, it neglects evolution in the attacked organism or system. Fourth, it pays insufficient attention to war-termination issues. Fifth, the current model neglects the reality of the challenges posed by the post-attack damage assessment architecture, the Air Tasking Order (ATO) system, and the reality of combined arms operations. Sixth, the assertion that "information is the bolt that holds the rings together" seems to give the lie to the entire theory. And last, in the world that is emerging, there may be little room for this type of air campaign. Each of these lesser challenges deserves a few words.

The simultaneous attacks celebrated by the five-rings theorists in the Gulf War did *not* occur simultaneously. They occurred sequentially and over time. For airpower to be effective, air superiority, or control of the air, is necessary. To achieve air superiority, enemy air defenses must be defeated, circumvented, or suppressed. Thus, and even though the initial onslaught may have attacked other targets in other categories, elimination or reduction of the capabilities of the enemy's air defenses always must be the first priority of an air campaign.[27] If the theorists and air campaigners assert that suppression of enemy air defenses was not their first priority in time and in space, that assertion appears to contradict current air doctrine. If they accept that coping with enemy air defense capability was the first priority, but reply that the first wave of attacks included other targets, then the attacks were not simultaneous, but merely very close in time. (Zeno probably would argue that any separation in time, however, constitutes serial warfare.) In truth, the opening salvos of the Desert Storm air campaign were directed—as they must be for air power to be effective—against the enemy air defense system, the "crucial first step" in the air campaign.[28] Sea-launched Tomahawk cruise missiles and Army Apache helicopters were part of this first wave of airpower attacks for airpower's benefit. How quickly other targets in the series followed, becomes less relevant to the theory. A compressed serial attack is still serial warfare, even though time compression may create the appearance and, more important, the effect of simultaneity.[29]

Because the five-rings model for air campaign planning asserts that the consequence of its attacks will be paralysis of the enemy system, it in effect asserts that the Napoleonic and Clausewitzian "decisive battle" is its aim. Moreover, it seeks to annihilate enemy capability.[30] (It does this, by the way, even while some of its advocates suggest that their theories now might have rendered much or most of Clausewitz irrelevant.) If the aim of the air campaign is not achieved—that is, if the consequent is not affirmed—then the fault must reside not in the air campaign, but somewhere else.[31] Dogmatic adherence to the air campaign plan list of priority targets is necessary to "prove" the theory. Close air support, the theory holds, is less important than strategic attack. If sorties have to be reapportioned because of some "ground emergency," then the dogma has been violated and, of course, the opportunity to win a decisive battle may have then been lost. Where the targeting list is followed religiously, failure to achieve a decisive battle can also be attributed to inadequate intelligence. Or it could be bad weather, the bane of aviation. Or it could be caused by an adaptive enemy.

The reality is that organisms are autopoietic; that is, they struggle to preserve themselves.[32] Any attacked organism can be expected to struggle for survival by responding and adapting to stimuli, to internal changes, and to its new environment. Rigid adherence to an air campaign plan specifying a series of parallel attacks in advance is rigid adherence to a set of attacks designed against the initial organism, not the evolved one. The danger with a wonderfully deterministic air campaign plan is that it may adapt poorly to an organism that evolves in unexpected ways. When flying weather impedes mechanical execution of the air campaign plan, allowing the enemy respite and the opportunity to recover, it is the fault of chaos. When mobile missiles are introduced in unexpected ways, it is the fault of intelligence. When operations against mobile missiles deplete sorties intended to achieve the decisive victory of the air campaign, it is the fault of the politicians. Airpower advocates did in fact argue that the Iraqi Scuds were not militarily significant.[33] That the missiles might have rent the Gulf War Coalition asunder had they not been actively pursued and engaged shows an immature understanding of what constitutes "military" significance in state warfare. Mobile missiles ought to have been priority targets in the air campaign: as political weapons they might have altered the course and outcome of the war.

War termination issues, not neglected in the SIOP, also appear to be neglected in the five-rings approach. The posited aim of the air campaign is strategic paralysis—the expectation being that "paralysis" must somehow equate to "surrender." The reality is somewhat different. Wars may end because the losers sense that there is something they value more than the object of the war and that continuing the war imperils preservation of this more important value or preference set.[34] The five-rings model attacks everything but population centers, perhaps encouraging the enemy to fight to the death. Even simple attacks can then have unintended consequences. Attacks against communications designed to separate leadership from fielded forces, for example, may also deprive leadership of feedback regarding damage to the organism. Thus, the organism may neither realize its paralysis nor behave as a paralytic. Certainly, if attacks annihilate a large part of the enemy's capability, defeat in detail is then possible, whether the enemy fights on or not—that is, if public opinion on the winning side supports the bloodletting required to defeat an enemy in detail. None should, of course, underestimate the ruthlessness of a United States forced to fight for its vital interests. It would be equally foolhardy to underestimate the power of public opinion in fights not perceived by our citizens as involving *their* vital interests.

The five-rings approach may work well against a weak enemy and a transparent target set. If the enemy—the resistant element described by Clausewitz—is *not* weak or stupid, or if the target set is characterized by attention to ninjitsu, problems arise for the air campaign planner. Prewar intelligence, very effective damage assessment, and close coordination are all required to make the ATO system function effectively. (Where the ATO is less effective and the fault cannot be attributed to "intelligence," it is, we are told by some, the fault of the Army and the Marines.[35]) The ATO takes a long time to produce.[36] This fact alone ensures that it directs attacks against an organism that no longer exists. If damage assessment is imperfect—normally imperfection is the status quo unless the weather is perfect, intelligence is precise and abundant, and the enemy is perfectly inept or stupid—the problems are compounded. That Saddam was not stupid, albeit an excessively bold risk-taker (and even though Gen H. Norman Schwarzkopf reviled him as a strategist), is shown by his pawns gambit on the border of Kuwait in 1994. Perhaps the Gulf War educated him.

If the Gulf War educated us, we should now appreciate that warfare is a poor laboratory for validating air-only, or naval-only, or ground-only theories.[37] Warfare against a small and inferior state is an even poorer laboratory. We fight with combined arms and depend on their interaction, their combined effects, to defeat the enemy's strategy. An enemy facing a 400,000 or a one-half million person allied army on its border will invariably behave differently than one only facing air attacks, no matter how wonderful the pounding to which air subjects that enemy. It may be as divisive as it is short-sighted to resign forces in other media to the null set when attempting to use an actual fight with a third-, fourth-, or tenth-rate opponent to illustrate or prove a theory.

Airpower—Army, Navy, Marine Corps, Air Force, and other nation airpower—*was* powerful in Desert Storm. Of that there can be no doubt. But was it powerful because Iraq was so inferior? Was airpower powerful alone, or was it powerful because of the force—the well-armed, well-trained, belligerent and hot-blooded human beings—poised to take the fight to Baghdad on land and from the sea? Would not have true parallel war, horizontal and vertical parallel war, brought the interactive power of land warfare and amphibious assault to bear on Iraq even as the air campaign unfolded? Has our desire for few casualties become yet another weakness; a weakness leading us away from sound strategy?[38]

This last question is an important one when examining the air campaign's quest for parallel war. Wars occur and warfare occurs within a much broader context than the battlespace. Will the strategic context of the future—the entire social, political, economic, and military gamut of goals, interests, and behaviors—tolerate the kind of Desert Storm air campaign advocated? Parallel war is and has been a wonderful theory. Yet, the move from theory to practice is both a torturous and tortuous one. Preparedness to execute the SIOP, for example, cost the United States trillions of dollars over decades. Preparedness to execute a Desert Storm-type air campaign against any but small and weaker states might require an equivalent investment. Would such warfare work against a large country? Against a peer? Will the United States ever again have the surplus resources it had in Desert Storm? Not likely, seems to be the answer.

Iraq was and is a small country. When proportional silhouettes of Iraq are superimposed over a larger nation, as they are in Figure 3, the aerial achievements of Desert Storm appear in a different light. This is not to suggest any adversarial relationship with or hostile designs against China. Rather, this perspective merely illuminates the fact that Iraq is a very small country.

Finally, if information is "the bolt that holds the five rings together," then information is the decisive center of gravity. Accordingly, should we not aim all attacks at information? Even though our understanding of information operations or information warfare is imperfect and immature, the evolving

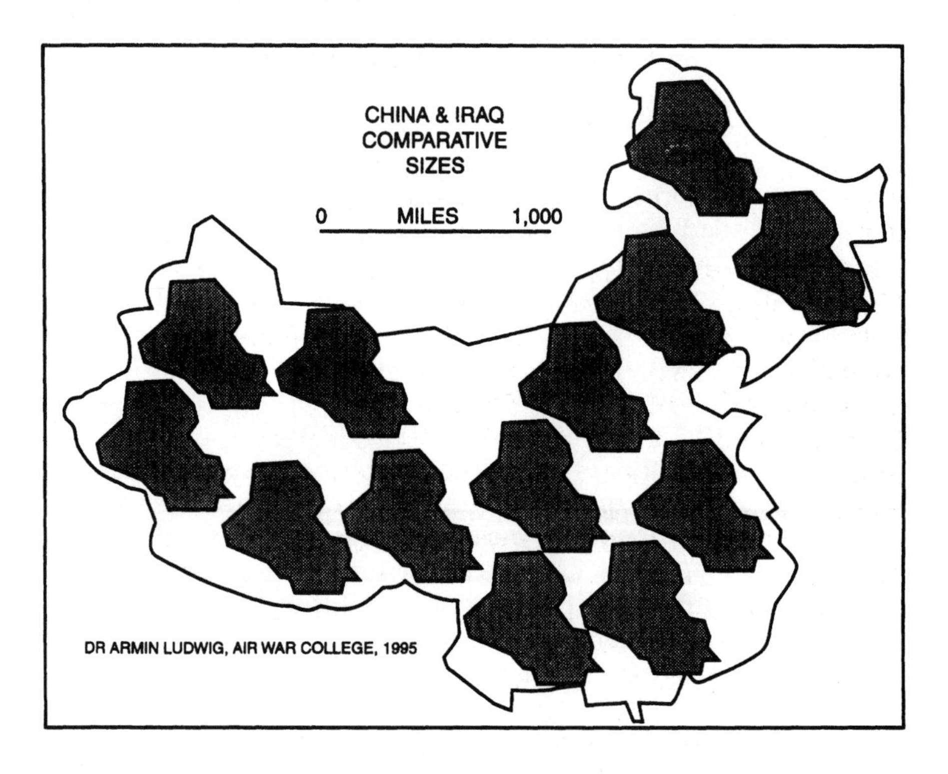

Figure 3. China and Iraq—Comparative Sizes

theory demands that the connectivity between and within the rings becomes the object of attack. Electrical power production facilities, roads and rails, airfields, missile production installations, and government buildings are not difficult for the air campaigner to target. Information is, or would be, difficult to target. Inclusion of information as a lucrative target may be a trendy afterthought on the part of airpower advocates on the one hand. On the other hand, it may be creation of another precondition for success of the air campaign. The precondition, if not met, then becomes the fault not of the theory, but of prehostility target intelligence or the clumsy execution of a brilliant air campaign plan.

Defeating Parallel War

Perhaps the more valuable contribution the five-rings model makes to the study of warfare is that it elucidates how one can falsify or defeat the theory. The first priority, the best way to defeat an adversary, Sun Tzu tells us, is to defeat an adversary's *strategy*. Air campaigners do not appear to be strategists. More likely they are air tacticians—their protestations to the contrary notwithstanding. If the five rings support the overall strategy or constitute the strategy of the air campaign, how can one defeat the strategy? There are at least five ways, with each one examined in turn. (Recall, please, that this discourse aims at theory and antithesis. In practice, some of these countervailing means are as risky as they are illegal. Even so, we would do well to keep in mind that our conceptions of risk and legality may be ours alone. There are antithetical notions out there in the real world.)

- Disguise, diversify, and demassify the system. Force the air campaign planner and the air campaigner to strike what appear to be widely distributed and primarily civilian targets. The center of gravity of United States armed forces may be public opinion. If public opinion is the Clausewitzian hub upon which all movement depends in most democratic states, attacking this hub is both a prewar and wartime priority. To defeat parallel war even before it commences, a wise adversary will strive to disguise, diversify, and "demassify" key elements of the system so that total war is necessary. Such a cunning adversary will have mobile systems wherever possible. Where mobility is not possible, the adversary will embed military activities in civilian ones. Weapons production will occur in facilities producing civilian goods. Weapons research activities will be collocated in hospitals, universities, and religious centers. Command and control transmitters and receivers will be placed on schools, hotels, temples, and recreational facilities. Airfields will be joint commercial-military facilities, routinely used by commercial and military entities. Dual-use systems—telecommunications media, fiber optics, direct broadcast and very small aperture satellites—will be used for administrative military communications. An adversary might build tanks in automobile factories, ballistic missiles in refrigerator factories, might commingle military and civilian transport, and could build its military garrisons in populous areas. The adversary would also be wise to move from military leadership hierarchies to military leadership networks. Where possible, the enemy will put foreign contractors in all militarily significant facilities. The adversary will encourage engagement, enlargement, tourism, and foreign investment. The adversary's objective is to up the ante for the attacker by forcing him to war on the innocents and against the investors. The wise adversary will attack and defeat its opponent's strategy even before declaration of a state of hostilities.

- Acquire weapons of mass destruction (WMD) on mobile systems. WMD so raise the risks and consequences of an attack that mere possession of WMD and mobile delivery systems may impose prewar paralysis on the attacker. To defeat parallel war, the adversary will distribute WMD among the innocents with the same impunity that other military capability is commingled. Nuclear weapons production facilities may have bold, bright signatures. Chemical and biological weapons production facilities do not. Since those may be the WMD of choice in the third wave, an adversary intending to defeat parallel war will acquire them. Acquisition of any deliverable WMD changes the political and military calculus dramatically.
- Where immobile, be invisible. One of the best ways to create invisibility is to go underground; the deeper, the better. Lateral is better than merely vertical. A cunning adversary will realize that whatever is the key node in the underground facility should also reside beneath a civilian facility. Schools, orphanages, and "baby milk" factories may have high utility for concealing these kinds of basements. Distinctive painting or marking, source-identifiable military communications, and even uniforms do not contribute to invisibility. The obvious and key point is that whatever cannot be identified cannot be targeted. An astute adversary will also capitalize on the attackers cultural myopia and mirror-imaging.

- When attacked, mutate. Defeating parallel war also requires planning how the organism will adapt, transform itself, and recover, if attacked. The adversary intent on defeating parallel war will anticipate a reduction in energy levels after an attack. Even so, the adversary also will realize that a planned mutation at the bifurcation point is superior to a random one. Especially cunning adversaries will choose an asymmetrical and unpredictable response to attacks, not a symmetrical or predictable one. If, for example, electrical power production comes under intense attack, the adversary might respond by intentionally shutting down all visible electrical power. (One cannot be especially cunning without having considered and prepared for this in advance.) This unexpected mutation makes damage assessment difficult. The air campaign planner may cope with this difficulty by forcing an extensive search for corroboration that attacks have achieved required damage expectancies, may fall into the trap of wishful thinking and reallocate sorties to other roles, or—and this is the likely case—may mindlessly continue to adhere to the installation-driven or target-driven air campaign plan. Another useful mutation might be to withdraw uniformed fielded forces and employ terrorists or special operations forces in the attacker's homeland.

- Attack information. The five rings exist in peacetime as well as in wartime. If information is indeed the bolt that holds them together, an adversary will realize that attacks should begin in the prehostility phase. The object of preliminary attacks in the prehostility phase will be to paralyze or destroy a target set called "any public opinion that does not support my aims." Combining propaganda with more active measures, such as assassination and other kinds of terrorism, may prevent a weak-willed attacker from taking the offensive. (On the other hand, active measures might stimulate uncharacteristically ferocious responses.) Failing the success of propaganda and anticipating an attack, an adversary may turn to "worms" and "viruses" as its swords. As one communications analyst has noted:

> Computer "hackers" have relatively easy access to software programs called "worms" and "viruses." A worm is used to delete a portion of a target computer's memory, and the aptly named virus calls forth a machine's files and copies itself onto them, creating an unfixable mess. Database records can be (and have been) altered by outside interference, just as broadcasting airwaves have been intercepted and preempted. Even on a small budget, it seems, where there is a will there is a way.[39]

As Brigadier V. K. Nair, VSM (Retired), wrote in his book *War In The Gulf: Lessons for the Third World:*

> Active measures to degrade attacking electronic systems should be cost effective and simple. For example, the most sophisticated system such as that of the United States, could be totally disrupted by the projection of a suitable virus that would automatically find their [sic] way back into the computers on which the systems are dependent. Cheap, simple and effective avenues must be exploited on a priority.[40]

Thus, it is only a matter of technique and time before these countermeasures to parallel war are employed. What are the counter-countermeasures? There may be none, although information technology may provide the homeopathy of future warfare.[41] In the quest for primacy in warfare, the pendulum will continue to swing between measures and countermeasures. The elusive search for the technologies, the weapons, and the concepts of operations and organization that allow *Vernichtungschlacht* likely will continue.

Conclusions

There may not really be much that is revolutionary in contemporary notions of parallel war and hyperwar. It is the logical evolution of a nonnuclear SIOP accelerated in serial applications. The ability to execute a nonnuclear SIOP against a large state or peer would require SIOP-level investments. After acquiring the capability, an adversary could checkmate it by simple tactical adjustments modulating the strategic environment. Dispersing and disguising targets and making the public opinion consequences of striking them unacceptable cause the more obvious of these modulations. Absent the powerful real or imagined survival motives that impelled the

SIOP, it is unlikely that the United States or any other nation will acquire the kind of airpower required to satisfy the needs of the parallel war and hyperwar air campaign. Absent the acquisition of the required technologies, the theory remains a rather large and important footnote to the Gulf War.

A problem with contemporary air campaign theories may be the progressive detachment of these presumptive airpower theories-presumptive from the realities of warfare.[42] The new theories seem to be less about warfare than they are about the ways in which some believe the battlespace ought to be apportioned *and* the resources that ought to be acquired once the roles of force elements are determined. Warfare is about human beings, human aspirations, and human passions. No one should thoughtfully relegate the study of warfare to the investigation of sterile technology and the targets that reside in precisely defined systems or rings. Warfare between humans is a hot thing, not a cold thing. It is more about blood, fear, surprise, and friction than it is about technology. Precision targeting depends on speed and certainty, including the assurance that humans can know and understand causal relationships. Parallel war theorists require the ready availability of the technologies that allow speed, precision, and the chimera of accurate knowledge and authentic understanding. Parallel war requires massive resource investments.[43]

Thus, airpower in this current formulation seems to have become no more nor less than the power of detached, dispassionate technology.[44] Technology is applied science. The science of the parallel war theorists is cold, deterministic, and—from the perspective of those who value jointness or integration—misapplied. "If this, then that" is a supposition that warfare rarely substantiates. The *hubris* of contemporary airpower theorists may be that they so badly want airpower to be dominant or decisive that they have made the Desert Storm air campaign the mechanistic template for all future wars. This want is a potentially dangerous weakness. Airpower, as Carl Builder suggests in *The Icarus Syndrome*, still lacks a theory.[45] Even so, one must applaud those who search for one, even while cautioning them that they may not have found one yet.

Notes

1. Michael R. Gordon and Bernard E. Trainor, *The Generals War: The Inside Story of the Conflict in the Gulf* (Boston: Little, Brown and Co., 1994); and John A. Warden III, "Air Theory for the Twenty-first Century," in Karl P. Magyar, ed., *Challenge and Response: Anticipating US Military Security Concerns* (Maxwell AFB, Ala.: Air University Press, 1994), 326–329. Warden asserts that one of the ten concepts that describe "the revolution of the Gulf War and must be taken into account "as we develop new force levels" is "The dominance of airpower." Some disagree regarding airpower's contribution in the Gulf War. See Robert H. Scales, *Certain Victory* (Washington, D.C.: US Government Printing Office, 1993).

2. "The function of the Army and the Navy in any future war will be to support the dominant air arm." See James H. Doolittle, speech to the Georgetown University Alumni Association, 30 April 1949, cited in Robert D. Heinl, Jr., *Dictionary of Military and Naval Quotations* (Annapolis: United States Naval Institute, 1966), 6.

3. Airpower advocates believe airpower was both dominant and decisive in the Gulf War. In the wake of the war, the principal architect of the Desert Storm air campaign plan, Col John A. Warden III, emerged as *the* lead airpower theorist in the United States Air Force. His prewar book, *The Air Campaign: Planning for Combat*, is pirated, translated, and studied abroad. The Swedish Air Force uses his theories as the model for the most stressful kind of air attack against which they must be prepared to defend. In Australia, Warden is ranked along with Douhet and Trenchard as an important air theorist. Colonel Warden is a former commandant of the Air Command and Staff College, Air University.

4. Carl von Clausewitz, *On War*, ed. and trans. by Michael Howard and Peter Paret (Princeton: Princeton University Press, 1976), bk. one, chap. 1, 79. Emphasis in original.

5. The characteristics of parallel war and the air campaign primarily are those provided by Colonel Warden. Like Col John R. Boyd before him, most of Warden's brilliant exposition of parallel air warfare is in the oral tradition: briefings and briefing slides. Where primary written works are available, they are cited.

6. Warden, "Air Theory for the Twenty-first Century," 311–318.

7. Dan Hughes reminded me that in physics, just as in *jujitsu*, a body has and can have only *one* center of gravity. Dr Hughes is a colleague at the Air War College, Air University.

8. Warden, "Air Theory," 326–331. See also John A. Warden III, "Employing Air Power in the Twenty-first Century," in Richard H. Schultz, Jr., and Robert L. Pfaltzgraff, Jr., eds., *The Future of Air Power in the Aftermath of the Gulf War* (Maxwell AFB, Ala.: Air University Press, 1992), 64–69.

9. John R. Pardo, Jr., "Parallel Warfare: Its Nature and Application," in Karl P. Magyar, ed., *Challenge and Response: Anticipating US Military Security Concerns* (Maxwell AFB, Ala.: Air University Press, 1994), 277–296.

10. David Deptula, draft, unpublished manuscript, "Firing for Effect—Change in the Nature of Warfare," 14 October 1994, 12–21. The author is grateful to Col Jeffrey Barnett, OSD/NA, for pointing out this essay.

11. "Information Dominance Edges Toward New Conflict Frontier," *Signal Magazine*, August 1994, 37–40. See also George Stein, "NetWar-CyberWar-Information War," June 1994, forthcoming in *Air Power Journal*; John Arquilla and David Ronfeldt, "Cyberwar is Coming!" *Comparative Strategy*, 2, April–June 1993, 141–65; and Norman B. Hutcherson, "Command and Control Warfare: Putting Another Tool in the War-Fighters Data Base" (Maxwell AFB, Ala.: Air University Press), 5 August 1994.

12. Conversation with Colonel Warden, 10 November 1994. See also Warden, "Air Theory," 329–330.

13. John Warden sometimes uses the example of 150 tornadoes simultaneously striking the United States. Such a large number of tornadoes hitting at the same time would make recovery exceedingly difficult since recovery resources could not be shared easily.

14. Jeffrey R. Cooper, *Another View of the Revolution in Military Affairs* (Carlisle Barracks, Pa.: US Army War College, 1994), 29–30. Cooper's insightful conclusions about the constraints on an authentic revolution in military affairs (RMA) can also be applied to the air campaign theory of parallel war.

15. The United States Strategic Bombing Survey, Summary Report (European War), 1945; reprinted in *The United States Strategic Bombing Surveys (European War) (Pacific War)* (Maxwell AFB, Ala.: Air University Press, 1987); and Thomas A. Kearney and Elliot A. Cohen, *Gulf War Air Power Survey: Summary Report* (Washington, D.C.: Government Printing Office, 1993). The bomber does not "always get through" (sometimes bombers have difficulty getting through the budget process), nor are precision weapons of much use unless supported by precise prestrike information on the target and poststrike damage to it.

16. J. C. Wylie, *Military Strategy: A General Theory of Power Control* (New Brunswick, N.J.: Rutgers University Press, 1967), 26. Italics added. Wylie introduced the differentiations of sequential and cumulative in "Reflections on the War in the Pacific," *US Naval Institute Proceedings* 78, no. 4, April 1952, 351–361. See also J. C. Wylie, *Military Strategy: A General Theory of Power Control*, in *Classics of Sea Power* (Annapolis: Naval Institute Press, 1989), 101.

17. Desmond Ball, "Development of the SIOP, 1960–1983," in Desmond Ball and Jeffrey Richelson, eds., *Strategic Nuclear Targeting* (Ithaca: Cornell University Press, 1986), 80–81.

18. A historian assigned to the Gulf War Airpower Survey team, speaking under the promise of nonattribution, pointed out to the Air Force officers detailed to assist with the study that the targeting logic of the Instant Thunder air campaign briefing was the same as the SIOP. He was told that "parallel war" was "an Air Force idea," while the SIOP was a "joint idea." Parallel war is an old idea. It remains an important one for nuclear operations, since deterring and fighting a peer may require a return to dependence on nuclear weapons.

19. Ballistic missile attack options evidenced extraordinary coherence and simultaneity. Theoretically it was possible, for example, to time sea-based and land-based ballistic missile launches so that all attacking warheads arrived at the first possible point of enemy radar detection

simultaneously. Source is a discussion with Lt Gen Jay W. Kelley, USAF, Maxwell AFB, Alabama, 1 December 1994. General Kelley is the commander of Air University.

20. What may be "new" here may be that some in the tactical air forces discovered the approach that strategic air and missile forces had long taken to warfare. The Air Force Directorate of Warfighting Concepts Development (AF/XOXW) lays claims to the notion of parallel warfare, using the differences between parallel and serial electrical circuits as the illustrative model. Admiral Wylie would probably be pleased; imitation is the highest form of flattery.

21. CMDR J. M. van Toal, OSD Net Assessment, suggests that the difference between employing nuclear weapons and nonnuclear weapons creates distinctions that are "definitive."

22. A state like Iraq is an even easier target set. A very senior air officer involved in the Desert Storm air campaign, speaking under the promise of nonattribution, asserted that achieving air supremacy against Iraq was about as difficult as "shooting at a tethered goat."

23. Dan Hughes suggests that there are at least two sides to this issue. Perhaps it was awareness of this likelihood in the leadership of the former Soviet Union (or the US) that made such a US (or Soviet) attack unnecessary.

24. Alvin and Heidi Toffler, *War and Anti-War: Survival at the Dawn of the 21st Century* (Boston: Little, Brown and Co., 1993). Desert Storm was not third wave warfare. It may, however, have been nearly the epitome of second wave warfare.

25. John Keegan, *A History of Warfare* (New York: Alfred A. Knopf, 1994); Martin van Creveld, *The Transformation of War* (New York: Free Press, 1991); and Carl Builder, "Guns or Butter: The Twilight of a Tradeoff?" (May 1994), a presentation to the USAF Air University National Security Forum, Maxwell AFB, Alabama. Used with permission.

26. Maj Gen Peter D. Robinson, former commandant of the Air War College, points out that the techniques of mobility, deception, disguise, dispersion, misinformation, and diversion are still useful ways of creating *ninjitsu*.

27. The demassification and dehumanization of attack systems could make the suppression of enemy air defenses a much lower priority. A strength of the SIOP was the land-launched and sea-launched ballistic missiles that preceded attacks by manned aircraft. SSBNs are "stealthy" by virtue of their operations. The five-rings' parallel war air campaign planner may not have such missiles to employ. Cruise missiles and stealth as a design feature become the analog of intercontinental ballistic missiles and were employed— as we would expect them to be if the SIOP and parallel warfare theories are the same—in advance of manned systems.

28. James A. Winnefeld, Preston Niblack, and Dana J. Johnson, *A League of Airmen: U.S. Air Power in the Gulf War* (Santa Monica, Calif.: RAND, 1994), 120–121.

29. Barry R. Schneider points out that "compacted serial warfare may 'be parallel warfare in effect " if the adversary does not have the time to respond effectively. I would argue that this further illuminates a limitation in the theory: to produce the "effect" desired in time, we must use Clausewitzian

mass in space. Absent nearly unlimited mass, the notion of parallel air warfare may only be valid if applied against small countries. Dr Schneider is a colleague at the Air War College, Air University.

30. Advocates argue that the objective is not annihilation but "control." The possibility that control actually is secured by annihilating military capability should not be overlooked.

31. David Deptula, draft, "Firing for Effect," 20.

32. Erich Jantsch, *The Self-Organizing Universe* (Oxford: Pergamon Press, 1980), 7, quoted in Margaret J. Wheatley, *Leadership and the New Science: Learning about Organization from an Orderly Universe* (San Francisco: Berrett-Koehler Publishers, Inc., 1992), 18.

33. Alexander S. Cochran et. al., *Gulf War Air Power Survey, Volume I: Planning* (Washington, D.C.: Government Printing Office, 1993), 103–4. Dr Cochran is a colleague at the Air War College, Air University.

34. Joseph A. Engelbrecht, "War Termination: Why Does A State Decide To Stop Fighting?" (Unpublished doctoral dissertation, Graduate School of Arts and Sciences, Columbia University, 1992). Col Engelbrecht is a colleague at the Air University's Air War College.

35. According to a senior Air Force officer speaking to the Air War College under the promise of nonattribution, the Army is to blame for its passion for "deep attack" and the way in which it uses the fire support coordination line (FSCL) to apportion the battlespace. The Marines are to blame for withholding sorties from the joint forces air component commander (JFACC). All of this, the official argued, reduces the power of airpower.

36. The importance of time and opportunity in warfare, dimensions currently neglected in Army doctrine, is developed in Robert R. Leonhard, *Fighting by Minutes: Time and the Art of War* (Westport, Conn.: Praeger Publishers, 1994). See the discussion of some of the limits of the ATO system on page 74.

37. As Gen Al Gray, the former commandant of the Marine Corps, says, "I've never seen a battlefield too crowded to exclude anyone who wants a shot at the enemy."

38. The Department of Defense plans to increase investment in nonlethal technologies significantly. One wonders if unwillingness to kill the enemy or employ friendly ground forces in mortal combat against an adversary's homeland forces renders the United States weak to the point of impotence. If the Army and Marine Corps ground forces are unused in combat, they will eventually become less useful for and effective in combat.

39. Joseph N. Pelton, *Future View: Communications Technology and Society in the 21st Century* (Boulder, Colo.: Baylin Publishing, 1992), 196.

40. V. K. Nair, *War In The Gulf: Lessons for the Third World* (New Delhi, India: Lancer International, 1991), 110.

41. Said another way, information warfare may provide the antidote for parallel warfare. Eventually, and embellished by technology to intrude into acoustic, tactile, olfactory, and visual space, information technology may be the antidote for all warfare. See Marshall and Eric McLuhan, *Laws of Media: The New Science* (Toronto: University of Toronto Press, 1988); Gregory L. Ulmer, *Heuretics: The Logic of Invention* (Baltimore: Johns Hopkins University Press, 1994); and Diane Chotikul, "The Soviet Theory of Reflexive

Control in Historical and Psychocultural Perspective: A Preliminary Study," Technical Report NPS55-86-013 (Monterey, Calif.: Naval Postgraduate School, 1986).

42. A theory that asserts, for example, that an air force "owns" responsibility for everything that transits the air is not a warfare theory. A theory that asserts, for example, that an air force "owns" responsibility for the "deep battle" or the "high battle" is not a warfare theory. These are roles and missions, or defense resource appropriation, theories. A theory, on the other hand, hypothesizes that the battlespace is seamless.

43. Where the maintenance and modernization of large land armies continue to sap national investments, the required technologies are less likely to be forthcoming. France, England, and Germany provide good examples. The "army" structurally and financially dominates the defense debate and budget. Thus, airpower is foredoomed to structural impotence in France, Germany, and England.

44. A senior Air Force officer speaking to the Air War College under the promise of nonattribution asserted that "without technology, there is no Air Force." Technology is a tool, a means to an end. When it becomes an end in itself, it may serve itself instead of warfare.

45. Carl Builder, *The Icarus Syndrome: The Role of Air Power Theory in the Evolution and Fate of U.S. Air Force* (New Brunswick, NJ.: Transaction Publishers, 1994).

Overview: Information Warfare Issues

Information warfare is emerging as a potent new element of strategy. "Info War" is not the same as intelligence operations, although it is clearly related to intelligence. As it is emerging in Defense Department thinking, information warfare is an attack on an adversary's entire information, command and control, and, indeed, decision-making system.

As Air Force Plans puts it: information warfare is "any action to deny, exploit, corrupt, or destroy the enemy's information and its functions; protecting ourselves against the actions; and exploiting our own information operations." Information warfare is directed at shrinking or interfering with the enemy's Observe, Orient, Decide, Act (OODA) loop while expanding and improving our own. Strategists now speak of information as a "strategic asset," and planners describe the objective of information war as "information dominance."

Dr George Stein places emerging thinking on information warfare into the context of the Tofflers wave theory: "As first wave wars were fought over land, and 'second wave' wars were fought over physical resources and productive capacity, the emerging 'third wave' wars will be for the access to and control of knowledge." Related to information war are "net war" and "cyberwar." Net war is information war waged largely through communications systems. The 1991 Gulf War exhibited it via the Coalition's attack on the whole spectrum of Saddam Hussein's information, propaganda, command and control, etc.

But it is to the realm of "info propaganda" where Dr Stein calls our particular attention—the emergence of techniques "combining live actors with computer generated video graphics," and "fictive simulators," and other information manipulation which creates "virtual realties" that could seriously threaten a state's control. Cyberwar is the operational extension of information war and net war—the tactical disruption, then domination, and perhaps even the reordering, of an enemy's decision-cycle. However, Stein cautions that whether cyberwar can actually "shape" the battlefield, or merely generate chaos remains to be seen.

All this has obvious implications for command and control warfare against an adversary. Governments which rely for their legitimacy on insulating their societies from reality look particularly vulnerable to information war. Sound strategy, then, employs info war as an offensive element of operations, designed to disrupt or end an adversary's communications and decision making. Stein cautions, however, that democratic societies may be particularly vulnerable to attack by adversaries using information war, especially its components net war and cyberwar. Their communications infrastructure is wide open to attack for their "domestic computer, communications, and information networks . . . are very vulnerable to penetration, manipulation, or even destruction by determined hackers."

The essay by Col McLendon is built around a critique of the historical evolution of information war. He uses the allied deception and crypt analysis during WWII, and the impact of information technology during the Gulf War, to illustrate the quantum leap from propaganda and disinformation during WWII to the systematic application of information warfare in the post-Cold War age.

The use of allied "Ultra" intercepts of Nazi war plans and operations was critical to allied success. Indeed, Ultra intercepts, which were never compromised, "provided the bulk of intelligence to the Allies during the war." As Supreme Court Justice Lewis Powell, who had worked with Ultra during WWII prior to launching his legal career, stated, "In no other war have commanding generals had the quality and extent of intelligence provided by Ultra." Ultra gave the British advance warning of the German attack on England and of U-boat operations against Atlantic convoys. Indeed, Churchill had to make numerous painful decisions *not* to defend Allied assets he knew were going to be attacked for fear of alerting the Germans to Allied prior knowledge of their plans. An example was the Luftwaffe raid on Coventry.

The Gulf War brought the use of information deception and information war to its zenith. US Army units used the NAVSTAR GPS at the tactical level to locate Iraqi units even in the midst of desert sand storms. GPS, writes McClendon, "was *the* capability that made possible the [Allies] 'left hook' used to defeat Saddam Hussein's armored divisions." The sheer information overload attendant to Coalition operations was mind boggling: 700,000 phone calls and 150,000 messages per day; successful deconfliction of over 35,000 different communications frequencies; AWACS aircraft controlling 2,240 air sorties per day—more than 90,000 during the war with no midair collisions.

The Gulf War experience has spurred recognition within the US Defense Department that an ability to achieve "information dominance" could represent a new era in strategy formulation. "Global dominance," writes former Vice Chairman of the JCS, Adm David E. Jeremiah, "will be achieved by those that most clearly understand the role of information and the power of knowledge that flows from it."

Chapter 6

Information War - Cyberwar - Netwar

George J. Stein

In Arthur Waley's *Three Ways of Thought in Ancient China*, Chuang Tzu tells the story of a simple gardener who was shown a new tool that promised to change gardening. He laughed scornfully and replied,

> I used to be told by my teacher that where there are cunning contrivances there will be cunning performances, and where there are cunning performances there will be cunning hearts. He in whose breast a cunning heart lies has blurred the pristine purity of his nature; he who has blurred the pristine purity of his nature has troubled the quiet of his soul, and with one who has troubled the quiet of his soul Tao will not dwell. It is not that I do not know about this invention; but that I should be ashamed to use it.[1]

Strategy, according to the Department of Defense, is the "art and science of developing and using political, economic, psychological, and military forces as necessary during peace and war, to afford the maximum support to policies, in order to increase the probabilities and favorable consequences of victory and to lessen the chances of defeat."[2] For most people, it is obvious that the political and economic aspects of the national security policies of the United States are developed by the national political authorities (e.g., the president and the Congress) and, in dealing with foreign states or groups, executed by the Departments of State, Commerce, Agriculture, etc.

Policies for developing and using military forces are formulated by the national political authorities and conveyed to the armed forces through the secretary of defense. Few, however, have paid much attention to just how and by whom psychological forces are to be developed to support national policies. More importantly: What are psychological forces? By whom will these forces be used? With what authority? To what ends?

New tools and technologies for communication have created the potential for a new form of psychological warfare to a degree imagined only in science fiction. This new form of warfare is known as "information warfare." When we come to know the Tao of such an invention as information warfare, we may find that we are ashamed to use it.

The futurists Alvin and Heidi Toffler have argued that the United States armed forces need to develop a "systematic, capstone concept of military knowledge strategy." Such a strategy would include clear doctrine, and a policy for how the armed forces will acquire, process, distribute, and project knowledge.[3]

Quoting from the "Memorandum of Policy No. 30" (6 May 1993) of the US Joint Chiefs of Staff, the Tofflers argue that the US military is expanding the concept of Information War to include psychological operations aimed at influencing the "emotions, motives, objective reasoning, and ultimately the behavior" of others. Such an expansion would mirror the evolution of traditional warfare toward Information War. It would also mirror the progrssive steps of generating wealth from agriculture and natural resources in much earlier times, to the nineteenth and early twentieth century emphasis on industrial production, to the present emphasis on generating information products as a major new source of income.

As "first wave" wars were fought for land and "second wave" wars were fought for control over productive capacity, the emerging "third wave" wars will be fought for control of knowledge. And, since "combat form" in any society follows the "wealth-creation form" of that society, wars of the future will be increasingly "information wars."

Currently, there is neither formal military doctrine nor official definitions of information warfare. Despite the computer jargon involved, the idea of information warfare has not only captured the attention of military analysts—it also poses important policy questions.[4]

Despite the lack of authoritative definition, "netwar" and "cyberwar" are emerging as key concepts in discussing Information War. Originally these ideas seem to have come from the science fiction community. Consider, for example, the thought-provoking future war suggested in Bruce Sterling's *Islands in the Net*.[5] More recently, the concepts of netwar and cyberwar have been developed by John Arquilla and David

Ronfeldt in their important essay, "Cyberwar is Coming!"[6] Their suggestions provide a thoughtful starting point for exploring the issues that surround "information war."

Netwar, according to them, is a societal-level ideational conflict waged in part through internetted modes of communication. That is, netwar is most likely to be a nation-against-nation strategic level conflict. Netwar is about ideas and epistemology—what is known and how it is known. It would be waged largely through a society's communication systems.

The target of netwar is the human mind. One could argue that certain aspects of the cold war had the characteristics of a dress rehearsal for future netwar. Consider, for example, Radio Free Europe, the Cominform, Agence France Presse, or the US Information Agency. But netwar may involve more than traditional state-to-state conflict. The emerging of nonstate political actors such as Greenpeace and Amnesty International, as well as survivalist militias or Islamic revivalists, all with easy access to worldwide computer networks for the exchange of information or the coordination of political pressure on a national or global basis, suggests that the governments may not be the only parties waging Information War.

At first glance, netwar may appear to be a new word for old-fashioned propaganda. It would be comforting to believe that the "tried and true" methods (and limitations) of propaganda still worked. And the Gulf War showed that both Saddam Hussein and the Alliance were still of the old school. The war contained many elements of classic propaganda: accusations of bombed baby-milk factories and stolen baby incubators, inflated rhetoric and inflated stakes of the conflict; the future of the new world order and "the mother of battles" for the future of Islam; and the classic "us or them" polarization in which "neutrality" or unenthusiastic support was decried.

One element of traditional propaganda was absent, however, while Saddam Hussein became the "new Hitler" and President Bush was the "Great Satan," there was little demonization or dehumanization of the opponent. Perhaps the multicultural nature of the American-led alliance precluded turning the Iraqi army into something subhuman. Indeed, there may have been a spark of netwar genius in treating the Islamic Iraqi soldiers as "brave men put into an impossible situation by a stupid leader." Under such conditions, there is no dishonor in surrendering. And there may have been a glimpse of future netwar—it is rumored that Baghdad Radio signed on one morning with "The Star-Spangled Banner."

Traditional propaganda was usually targeted to influence a mass audience. Contemporary technologies have the potential to customize propaganda. Anyone who has received individually targeted advertising from a company specializing in "niche" marketing has had a momentary shudder upon realizing that some private companies seem to know everything about our tastes and buying habits.

Contemporary databases and multiple channels for information transmission have created the opportunity for custom-tailored netwar attacks. Computer bulletin boards, cellular telephones, video cameras tied to fax machines—all provide entry points and dissemination networks for customized assault.

A major new factor in information war results directly from the worldwide infosphere of television and broadcast news. Many people have begun to realize that governmental decisions are becoming increasingly reactive to a "fictive" universe created by, CNN and its various international competitors. This media-created universe is dubbed "fictive" rather than "fictional" because while what is shown may be "true," it is just not the whole, relevant, or contextual truth. And, of course, the close etymological relationship between "fictive" and "fictional" suggests how easy it is to manipulate the message.

Nevertheless, this fictive universe becomes the politically relevant universe in societies in which the government or its military is supposed to "do something." Somalia gets in the news and the United States gets into Somalia despite the reality of equally disastrous starvation, disorder, and rapine right next door in Sudan. There were no reporters with "skylink" in Sudan because the government of Sudan issued no visas. The potential for governments, parties in a civil war such as Bosnia, rebels in Chiapis, or even nonstate interests to manipulate the multimedia, multisource fictive universe to "wage societal-level ideational conflicts" should be obvious.[7]

Fictive or fictional operational environments, then, whether mass-targeted or niche-targeted, can be generated, transmitted, distributed, or broadcast by governments or all sorts of players through increasingly diversified networks. The niche-manipulation potential available to states or private interests with access to the universe of internetted communications such as the networks over which business, commercial, or banking information are transmitted to suggest that "Mexico" is about to devalue the peso could easily provoke financial chaos. The target state would not know what had happened until too late.[8]

Direct satellite broadcast to selected cable systems, analogous to central control of pay-per-view programs, again offers the potential for people in one province or region of a targeted state to discover that the maximum leader has decided to purge their clansmen from the army. To put it in the jargon of the infowarriors, info-niche attack in an increasingly multisource fictive universe offers unlimited potential for societal-level netwar.

Pictures Worth A Thousand Tanks

When the new, but already well-understood, simulation technologies of the Tekwar and MTV generation are added to the arsenal of netwar, a genuinely revolutionary transformation of propaganda and warfare becomes possible. Traditional propaganda might have attempted to discredit an adversary's news media showing, for example, that as the official casualty figures were demonstrably false, all "news" from the government was equally false. The credibility of the opponent was the target and the strategic intention was to separate the government from the people.

Today, the mastery of the techniques of combining live actors with computer-generated video graphics can easily create a "virtual" news conference, summit meeting, or perhaps even a battle which exists in "effects" though not in fact. Stored video images can be recombined endlessly to produce any effect chosen. Now, perhaps, "pictures" will be worth a thousand tanks.

Of course, "truth will out" eventually, but by the time the people of the targeted nation discover that the nationwide broadcast of the conversation between the maximum leader and "Jimmy Carter" in which all loyal citizens were told to cease fighting and return to their homes was created in Hollywood or Langley, the war may be over. Netwar is beginning to enter the zone of illusion.

This is not science fiction; these are the capabilities of existing or rapidly emerging technologies. Here's how it might work: through hitching a ride on an unsuspecting commercial satellite, a "fictive simulation" is broadcast. Simultaneously, various "info-niches" in the target state are accessed via "the net." These info-niche targets, and the information they receive, are tailored to the strategic needs of the moment: some receive reinforcement for the fictive simulation; other receive the "real" truth; others receive merely slight variations. What is happening here?

This kind of manipulation elevates the strategic potential of infopropaganda to new heights. This is not traditional propaganda in which the target is discredited as a source of reliable information. Rather, the very possibility of "truth" is being replaced with "virtual reality"; that is, "information" which produces effects independent of its physical reality. What is being attacked in a strategic level netwar are not only the emotions, or motives, or beliefs of the target population, but the very power of objective reasoning: this threatens the very possibility of state control.

Let us return to the previous scenario to play out its effects. The fictive simulation of the maximum leader's call to stop fighting would, of course, be followed immediately by a "real" broadcast in which state "Voice and Vision" exposes the netwar attack as propaganda invented by "culture destroyers in Hollywood." "Jimmy Carter" is denounced as a hoax. But the damage has already been done: it is all but impossible for the television viewers of the targeted state to tell which broadcast is true and which fiction, at least in a timely manner. In a society under assault across its entire infosphere, it will become increasingly difficult for members of that society to verify internally the truth or accuracy of anything. Objective reasoning is threatened.

At the strategic level, the ability to "observe" is flooded by contradictory information and data; more importantly, the ability to "orient" is weakened by the assault on the very possibility of objective reasoning; "decisions" respond increasingly to a fictive or virtual universe and, of course, governmental or military "actions" become increasingly chaotic as there is no "rational" relationship of means to ends.

It would seem, then, that strategic-level netwar or information war brings us within sight of that elusive "acme of skill" wherein the enemy is subdued without killing by attacking his ability to form a coherent strategy.[9]

Reality, however, may be far more complex than the infowarriors yet imagine, and victory not so neat. The idea of "societal-level ideational conflict" may need to be considered with all the care given to the conduct of nuclear war, as the "end state" of netwar may not be bloodless surrender but total disruption of the targeted society. Victory may be too costly as the cost may be truth itself.

What Is Truth?

Any discussion of information warfare, netwar, cyberwar, or even perception manipulation as a component of command and control warfare by the armed forces of the United States at the strategic level must occur in the context of the moral nature of communication in a pluralistic, secular, democratic society. That is, the question must be raised whether using the techniques of information warfare at the strategic level is compatible with American purposes and principles.

Likewise, the question must be raised whether the armed forces of the United States have either the moral or legal authority and, more importantly, the practical ability to develop and deploy the techniques of information warfare at the strategic level in a prudent and practical manner. There are good reasons to be skeptical.

According to the philosopher Eric Voegelin, the moral basis of communication in any society can be discussed in terms of its substantive, pragmatic, and intoxicant functions.[10] The substantive purpose of communication is the building or developing of the individual human personality; it is simultaneously the process by which a substantive, real-world community of "like-minded" persons is created, developed and sustained. Simply, it is the glue which binds a society together.

At the most trivial level, the moral purpose of substantive communication can be seen in contemporary American efforts to remove sexist or racist language from accepted use. At a more serious level, the debates in American society about prayer in the public schools illustrate a recognition of the substantive and formative nature of communication in society, as "private religious views," in the view of many, must not corrupt the public school formation of character for life in pluralistic, modern America.

Finally, any real world society rests on the substantive communication and understanding among its members. Again, in Voegelin's terms, society is no mere external structure of relationships; it is a "cosmion," a universe of meaning "illuminated with meaning from within by the human beings who continuously create and bear it as the mode and condition of their self-realization."[11]

The efforts of several nations such as China, Iran, or Saudi Arabia to insulate their societies from the effects of the global communications network illustrate their awareness that their cultures and societies may depend on a shared, substantive universe of discourse distinctive to their societies.

Even within the West, the French believe the continued existence of France as a distinctive society organized for action in history may require state intervention in the substantive content of communication within society.[12] That France seeks to limit the percentage of foreign broadcast material and American films in Europe illustrates the seriousness with which they consider the substantive nature of communication.

Voegelin's second construct, identifying the pragmatic function of communication in society, is reasonably straightforward. Pragmatic communication is defined by its goal and consists of the universe of techniques designed to influence other persons to behave in ways the communicator wishes. Only behavior matters. Most political and commercial communication is merely pragmatic. It is usually indifferent to the substantive moral content of the communication and intends to mold perception, and consequently behavior, to the purposes of the communicator. This pragmatic use of communication as an attempt at perception manipulation is, of course, the central essence of information war. Its use by the government and the armed forces is, consequently, the real issue.

Finally, the intoxicant function of communication in American society is equally straightforward. The addiction of a considerable part of the citizenry to talk shows, soap operas, romance novels, professional sports broadcasts, high-profile legal trials and other well-known forms of distraction and diversion is well catered to by the entertainment industry.

For Voegelin then, civil communication or public discourse in contemporary American society is dominated almost entirely by the intoxicant and pragmatic modes. More importantly, the absence of substantive communication in public life is defended by much of the secular and liberal political class in the name of freedom, pluralism, and multiculturalism.

Pluralistic America is supposed to be a society in which the formation of character or opinion is left, through the use of various means of communication, to private initiative. Government attempts at "communication" in an information war, especially if prosecuted by the armed forces, would raise serious questions in a pluralistic, multicultural society.

The official military view of strategy, recall, is the "art and science of developing and using political, economic, psychological, and military forces as necessary during peace and war to afford the maximum support to policies, in order to increase the probabilities and favorable consequences of victory and to lessen the chances of defeat."[13]

Strategy is the means to achieve an end, with military strategy serving political or policy purposes. A slightly different view of strategy, however, may highlight a problem of Information War. If strategy were seen as "a plan of action designed to achieve some end; a purpose together with a system of measures for its accomplishment," the limitations of infowar thinking are obvious.[14]

Sound military strategy requires influencing the adversary decision maker in some way that is not only advantageous but reasonably predictable. The goal is control, not chaos. A national security strategy of information war or netwar at the strategic level—that is, "societal-level ideational conflict waged in part through internetted modes of communication"—and an operational-level cyberwar or command-and-control warfare campaign to decapitate the enemy's command structure from its body of troops may or may not be "advantageous" but, more importantly, is unlikely to produce effects that are reasonably predictable.

Conflict is about a determinate something, not an indeterminate anything. If the goal of influencing the adversary's ability to "observe" by flooding him with corrupted or contradictory information and data; disrupting his ability to "orient" by the elimination of the possibility of objective reasoning; and forcing his "decisions" to respond to a fictive or virtual universe, "actions" will, of course, be produced, but they may well be actions which are chaotic, random, nonlinear and inherently unpredictable by our side as there is no "rational" relationship of means to ends.

In the context of military operational-level cyberwar or command-and-control warfare, this appeals to the infowarrior an attractive military strategy. The inherently unpredictable nature of combat, the notorious "fog and friction" of real battle, will be amplified for the enemy in a successful cyberwar.

A successful cyber-strategy depends on the ability of the local military commander to deploy his power assets, especially his combat forces, not merely to dominate the enemy decision cycle (which, after all, has just been rendered chaotic), but to exploit opportunities as they evolve unpredictably from the disoriented, decapitated, or irrational enemy actions. Whether, then, command-and-control warfare can "shape" the battlefield or will merely generate chaos remains to be seen.

Cyber-strategy is the control of the evolution of the battlefield or theater power distribution to impose the allied commander's "order" on the enemy's "chaos." As Sun-Tzu observed, "Those who are able to adapt to changes in the enemy and achieve victory are considered supreme."[15] The threat exists, however, that the destruction of enemy rationality may collapse "battle" into mere "fighting" with no outcome but surrender or death. Merely defeating hostile fielded military forces may be insufficient.

Sun-Tzu also observed that "when battles gain victories and attacks achieve occupations, yet these successes are not followed up, it is disastrous. This is known as 'persisting turmoil'."[16] Whether the recent Gulf War was a strategic victory or mere "battle" remains for historians to judge. Operational-level cyberwar may, then, be that very "acme of skill" which reduces the enemy will without killing. On the other hand, it may also be the abolition of strategy as it attacks the very rationality the enemy requires to decide for war termination.

Strategic Implications

The tools, techniques and strategy for cyberwar will be developed and, during wartime, should be employed. In many ways, cyberwar is more demanding than netwar.

But the resources, organization, and training needed for cyberwar will be provided once its war-winning, and casualty-reducing, potential is grasped by the national political leadership. Such a development would certainly be prudent. On the other hand, many of the tools and techniques of battlefield cyberwar can be applied to netwar or strategic-level information war. This application may not be prudent, however, as there are serious reasons to doubt the ability of the United States to prosecute information war successfully.

One reason is that the United States is an open society; it may be too vulnerable to engage in netwar with an adversary prepared to "fight back."[17] The communications infrastructure, the "information highway," is "wide open" in our society. American society may be terribly vulnerable to a strategic netwar attack; getting us to believe fictive claims appears to be what commercial and political advertising are all about, and they seem to be effective. Also we may find physical control and security to be impossible. The domestic computer, communication, and information networks essential for the daily functioning of American society are very vulnerable to penetration and manipulation—even destruction—by determined hackers.[18] In the future, these may not be amateurs but well-paid "network ninjas" inserting the latest French, Iranian, or Chinese virus into Compuserve or other parts of the internet.[19]

A strategic information warfare attack on America's communication systems, including our military communication systems, air traffic control system, financial net, fuel pipeline pumping software,and computer-based clock/timing systems, could result in societal paralysis.

Currently, for example, over 14,000 Internet databases are being used by over 30 million people in over 90 nations. Over 1,600 software pirates are prowling the Internet, some in the employ of hostile commercial or intelligence services. The recent "spy flap" between France and the United States over alleged US attempts to gather data on French Telecom may be indicative of the future.[20]

Infosphere dominance—controlling the world of information exchange—may be as complex and elusive as "escalation dominance" appeared to be in nuclear strategy.[21] It will certainly be expensive: the US business community and the US armed forces are required to devote ever more resources and attention to computer, communications, and database security. The resources and skills required for battlefield cyberwar are not insignificant, but the resources and skills required to wage Information War at the national strategic level would be massive.

The second reason to doubt US ability to prosecute an information war is that the political and legal issues surrounding info war are murky. What of congressional oversight? Would one "declare" information war in response, say, to an Iranian-originated computer virus assault on the FBI's central terrorist database? And what about preparing for it? How should we develop and implement a national capability for netwar?

While theoretically a requirement to develop or implement a national information war strategy, analogous to the nuclear-era single integrated operations plan, could be communicated from the president to the executive branch agencies, it is unclear whether there would be adequate congressional oversight. Which committees of the House or Senate would have control and oversight of policies attendant to information war, and which would have the power to inquire into the judgment of a local ambassador or military commander who wished to use the tools of cyberwar for a perception manipulation in peacetime that would shape the potential wartime environment?[22]

The US armed forces only execute the national military strategy—they do not control it. However, they are developing, quite appropriately, the tools and techniques to execute the national military strategy for operational-level cyberwar. They are simultaneously, albeit unintentionally, developing the tools and capabilities to execute a national strategic information war strategy. The former is their job under the Constitution; the latter may not be. Congressional oversight in the development of a national strategic-level information war capability is even more essential than oversight of the intelligence community.

The third reason to doubt US capabilities in prosecuting an effective information war is that such a "societal-level ideational conflict waged in part through internetted modes of communication" may simply be beyond the competence of the executive agencies that would have to determine the substantive content to be communicated. Pluralism is a great strength of American society, but perhaps a drawback in waging information war.

While diversity may make the formation and execution of domestic and even foreign policy more complex, the lack of a moral center or public philosophy in American society could render the political leadership incapable of building a consensus on strategic-level information war policies. And, since there is no single view of what is morally acceptable, but simply a host of contending views, a national security strategy of information war could be developed by the national security decision makers that lacked a moral consensus.

The technological wizardry does not change the humanity of the target. Unless the goal of information war is merely to unhinge people from their ability to reason objectively, and thereby create an interesting problem for post-conflict reconstruction, any strategic-level netwar or information war would seem to require the ability to communicate a replacement for the discredited content of the target society.

If, say, an information war were to be mounted against China to disrupt its drive for regional hegemony, the goal would be to "withdraw the Mandate of Heaven" from the rulers and "influence" the Chinese leaders and people to adopt the policies or behavior we find appropriate.

Put in terms of such a concrete policy goal, the philosophically problematic nature of information war becomes outrageously obvious. Does anyone really believe that the US national executive agencies, including the armed forces and the Central Intelligence Agency, know the substantive discourse of China sufficiently well to withdraw the Mandate of Heaven?

The final reason, then, can be stated in the form of a question: does anyone really believe that anyone in the US government has the philosophical sophistication to project an alternative discourse to replace the emotions, motives, reasoning, and behavior grounded in the Chinese reality we propose to influence? Would our "fictive" creation really have "virtual" effects. We might be able to use the armed forces or the CIA to destroy China's objective reasoning through a "successful" information war. Indeed, we might be able to loose anarchy in a society, but that is not usually the political goal of war.

Second Thoughts

The techniques being developed by the armed forces for a more narrowly constrained operational-level cyberwar was demonstrated in the Gulf War. Translated to the strategic level, however, netwar or information war is not a prudent national security or military strategy for the simple reason that neither the armed forces nor any other instruments of national power have the ability to exploit an adversary's society in a way that promises either advantageous or predictable results.

"Societal-level ideational conflict" must be considered with all the care given to the conduct of nuclear war, as the "end state" of a netwar may be total disruption of the targeted society. Conflict resolution, including ending wars this side of blasting people into unconditional surrender, assumes and requires some rationality—even if that rationality is the mere coordination of ends with means.

Moral reasoning and substantive communication may not be required; minimal reasoning and pragmatic communication *are* required. However, a successful all-out strategic-level information war may, however, have destroyed the enemy's ability to know anything with certainty and, thereby, his capacity for minimal reasoning or pragmatic communication.

In some exercises during the cold war "decapitation" of the Soviet military leadership in a hypothetical nuclear exchange was intended to defend the United States by preventing an escalatory or exploitative strike, nuclear or otherwise. Precisely how war termination would have been accomplished without an effective leadership will remain, hopefully, one of the great mysteries. The "decapitation" of the leadership is, however, often proposed as a key goal of an information war. That is, the credibility and legitimacy—even the physical ability to communicate—of the decisionmakers will be compromised or destroyed relative to their own population and in terms of their own worldview. And even if we merely "seize" his communication system electronically and substitute our "reality" into his society, with whom, then, do we negotiate the end of the conflict?

What confidence do we have that a call to surrender, even if communicated to the people by either the enemy leadership or our "net warriors," would be accepted as "real" and not another "virtual" event? And, depending on the content, intensity, and "totality" of a strategic information war, personalities could be flooded with irrational or unconsciousness factors—the clinical consequence of which is generally acute psychosis. How do we accomplish conflict resolution, war termination, or postconflict reconstruction with a population or leadership whose "objective reasoning" has been compromised?

Just as the mutually destructive effects of nuclear war were disproportionate to the goals of almost any imaginable conflict, so may be the mutually destructive effects of a "total" information war exchange on the publics exposed and subsequent rational communication between the sides. And as the techniques of "cyberstrike" proliferate throughout the world, enabling small powers, nonstate actors, or even terrorist hackers to do massive damage to the United States, "mutually assured cyberdestruction" may result in a kind of infowar deterrence. As Sun-Tzu advised, "without advantage, do not act; without gain, do not utilize; without crises, do not battle."[23]

Information War, then, may be the central national security issue of the twenty-first century. Therefore, the United States must develop a coherent national-level policy on the military and strategic use of new information warfare technologies. To facilitate this objective, the US armed forces are developing, under the rubric of command and control warfare, the technologies and systems that will provide the capability for "cyberwar."

It may be possible to control and exploit information so as to purposely generate stochastic chaos, though there are some doubts.[24] Many of the same technologies and systems can be used to develop a national-level capability for strategic "netwar." Here, however, there are genuine doubts. As Voegelin feared, it may not be possible to control and exploit information and information technologies to impose "a form on the remnants of societies no longer capable of self-organization" because their substantive universe of meaning has been destroyed or corrupted.[25]

Few info-warriors would claim the ability to "reorient" the former Soviet Union into a liberal society, or to influence the far more ancient barbarism in that heart of darkness, Rwanda. Perhaps strategic-level information war is, indeed, like nuclear war: the capability is required for deterrence; its employment, the folly of mutually assured destruction. But if the United States is to develop the capacity for information war, in the sure and certain knowledge that the technologies have already "proliferated" to both state and nonstate potential rivals, a realistic national consensus must be built.

It is useless to pretend that the proliferation of these technologies will not provide capabilities that can do serious harm. It is useless to pretend that military-based command and control warfare capabilities will not be developed, and it is useless to pretend that cyberwar technologies could not be turned to netwar applications. It is almost universally agreed that these capabilities are essential on the contemporary battlefield.

It is essential, then, that the president and the Congress give serious and sustained attention to cyberwar, netwar, and information war.

Notes

1. Arthur Waley, *Three Ways of Thought in Ancient China* (New York: Doubleday, 1939), 70.

2. Joint Pub 1-02, *Department of Defense Dictionary of Military and Associated Terms* (Washington, DC: US Government Printing Office, 1989), 350.

3. Alvin & Heidi Toffler, *War and Antiwar: Survival at the Dawn of the 21st Century* (Boston: Little, Brown & Co., 1993), 141.

4. The vocabulary of information warfare includes information war, information-based war, command and control warfare, information operations, C^3I, electronic warfare, and, in Russian usage, sixth-generation warfare.

5. Bruce Sterling, *Islands in the Net* (New York: Ace, 1988).

6. John Arquilla & David Ronfeldt, "Cyberwar is Coming!," *Comparative Strategy* 12: no.2 (April–June, 1993), 141–165.

7. Historians may record that Ecuador's posting of government communiqués on the internet at the beginning of the recent "war" with Peru may have been the first "netstrike." "A Borderless Dispute," *Newsweek*, 20 February 1995, 20.

8. H.D. Arnold et. al, "Targeting Financial Systems as Centers of Gravity: 'Low Intensity' to 'No Intensity Conflict,' *Defense Analysis*, 10, no.2, August, 1994, 181–208.

9. Ralph D. Sawyer, trans., *Sun-tzu: The Art of War* (New York: Barnes & Noble, 1994), 177.

10. Eric Voegelin, "Necessary Moral Bases for Communication in a Democracy," *Problems of Communication in a Pluralistic Society* (Milwaukee: Marquette University Press, 1956), 53–68.

11. Eric Voegelin, *The New Science of Politics*, (Chicago: The University of Chicago Press, 1952), 27.

12. John Andrews, "Culture Wars," *Wired*, May 1995, 130–138.

13. Rear Admiral J.C. Wylie, *Military Strategy: A General Theory of Power Control* (Annapolis, Md.: Naval Institute Press, 1967), 14.

14. Sun-Tzu, *The Art of War*, trans., J.H. Huang. (New York: Quill, 1993), 68.

15. Ibid, 109.

16. Peter Black, "Soft Kill: fighting infrastructure wars in the 21st century," *Wired*, July/August 1993, 49–50.

17. Paul Wallich, "A Rogue's Routing," *Scientific American* 272, no. 5, (May 1995), 31.

18. Winn Schwartau, *Information Warfare: Chaos on the Electronic Superhighway* (New York: Thunders Mountain Press, 1994).

19. Jean Pichot-Duclos, "Toward a French 'Economic Intelligence' Model," *Defense Nationale*, Jan 1994, 73–85, in *Federal Broadcast Information Service - West Europe*, 25 January 1994, 26–31.

20. John Arquilla, "The Strategic Implications of Information Dominance," *Strategic Review*, Summer, 1994, 24–30.

21. Joint Chiefs of Staff Memorandum of Policy 30, *Command and Control Warfare*, 8 March 1993.

22. Sun Tzu, 110.
23. Jeffrey R. Cooper, *Another View of the Revolution in Military Affairs* (Carlisle Barracks, Pa: Strategic Studies Institute, Army War College, 1994).
24. Eric Voegelin, "The Ecumenic Age," in *Order and History* (Baton Rouge: Louisiana State University Press, 1974), 117.
25. Brig V.K. Nair, *War in the Gulf: Lessons for the Third World* (New Delhi: Lancer International, 1991).

Chapter 7

Information Warfare: Impacts and Concerns

Col James W. McLendon, USAF

Information has always been a critical factor in war. Clausewitz said "imperfect knowledge of the situation . . .can bring military action to a standstill," and Sun Tzu indicated information is inherent in war fighting. Information warfare embodies the impact of information on military operations.

The computer age gives us the capability to absorb, evaluate, use, transmit, and exchange large volumes of information at high speeds to multiple recipients simultaneously. Multiple sources of data can be correlated faster than ever. Thus, the value of information to the war fighter has been magnified to a new level.

Churchill used information warfare when he used the Enigma machine to read German codes during World War II. He also used information warfare through his elaborate network emanating from the London Controlling Section, for its time a very complex intelligence and deception operation.

Lessons from Desert Storm gave impetus to this fourth dimension of warfare. It was in this conflict that the computer came of age, and presented us with new challenges, both offensively and defensively, that must be faced in the future. Not only do we have opportunities to enhance our offensive capabilities manyfold, but we must consider the additional vulnerabilities to our systems that come with this added capability. The widespread availability of information technology dictates that we carefully assess the vulnerabilities of the systems we employ.

Information warfare adds a fourth dimension of warfare to those of air, land, and sea. In this new dimension, we must stay ahead.

Information Warfare: Old Concept, New Technology

Given the wide realm of activities that might be included under the heading of information warfare, one might conclude that it is not a new concept but rather one that can be more aggressively employed today with new technology. Had the term *information warfare* existed in Churchill's day, he might have used it to describe his activities involving Ultra. Given the availability of communications and computer technology today, the potential for information warfare seems limitless. Unlike nuclear weaponry, however, this technology is not limited to a few nations. It is widespread and available to any country, and, in most cases, to any individual or group that wants it. It is for this reason that our pursuit of an offensive information warfare capability must not overshadow our appreciation of the need for a defensive capability.

This essay offers evidence of the need for a rigorous defensive information warfare capability. It includes a case study from World War II that demonstrates Churchill's creativity in using information warfare against the Germans and proposes that history may not have completely documented his activities in this endeavor. From World War II, we move to the Persian Gulf War, where information technology was embedded in virtually every aspect of Coalition operations. Our dependence on information media during the Gulf War is evident. This dependency may also equate to yet unknown vulnerabilities, thus highlighting the need for the protection of these media.

Information has always been a critical factor in war. According to Clausewitz, "imperfect knowledge of the situation . . .can bring military action to a standstill."[1] Pick up any book on war, and the value of information becomes clear. As indicated by Sun Tzu in 500 B.C., it is inherent in warfighting.[2] It may be obvious that the more an army knows about itself and its enemy, the stronger it will be in battle. What is not so obvious are the uses that may be made of information, and how knowledge can be manipulated to reinforce the strength of an army many times over.

Information warfare embodies the impact of information, or knowledge, on military operations. It is defined as "any action to deny, exploit, corrupt, or destroy the enemy's information and its functions; protecting ourselves against those actions; and exploiting our own information operations."[3] Additionally in this context, information warfare "views itself as both separate realm and lucrative target."[4] While this definition is new, the concept isn't. It is only as we come to terms with the benefits of the computer age that we realize the potential in conducting the operations described above.

The computer age gives us the capability to absorb, evaluate, use, transmit, and exchange large volumes of information at high speeds to multiple recipients simultaneously. Multiple sources of data can be correlated faster than ever. Until recently, masses of information were transmitted in the literal, or alphanumeric format, and had to be read and manually manipulated to be of any use. This made it difficult to sort the critical from the useful, and much of the information went into the burn bag. Today, much of that same information is transmitted to the war fighter digitally and presented graphically. Little goes to waste. Thus, the value of information, its uses, and our dependence on it have been magnified to a new level.

Duane Andrews, former assistant secretary of defense for C^3I (command, control, communications, and intelligence), describes information today as a "strategic asset."[5] The Tofflers go even further. In their discussion on third wave war, they refer to "knowledge warriors," describing them as "intellectuals in and out of uniform dedicated to the idea that knowledge can win, or prevent, wars."[6]

Maj Gen Kenneth Minihan, Air Force assistant chief of staff for intelligence, describes information warfare in more objective terms, which he says is really "information dominance."[7] In describing information dominance, he puts it this way:

> Information dominance is not "my pile of information is bigger than yours" in some sort of linear sense. It is not just a way to reduce the fog of war on our side or thicken it on the enemy's side. It is not analysis of yesterday's events, although proper application of historical analysis is important to gaining information dominance. It is something that is battled for, like air superiority. It is a way of increasing our capabilities by using that information to make right decisions, (and) apply them faster than the enemy can. It is a way to alter the enemy's entire perception of reality. It is a method of using all information at our disposal to predict (and affect) what happens tomorrow before the enemy even jumps out of bed and thinks about what to do today."[8]

The Navy presents the bottom line view: "Information, in all its forms, is the keystone to success."[9]

The Department of Defense and all of the services are doing more than paying lip service to this new dimension. In addition to their attempts to fund extended programs in this subject, senior military leaders are taking strong positions in favor of this capability. Unfortunately, while the United States holds the lead in information technology today, other nations, including developing nations, are rapidly gaining access to this capability. This is cause for concern, and the answers are not simple.

Information Warfare in World War II: How Far Did Churchill Go?

World War II saw many firsts. Some of the more significant examples were: large-scale air-to-air combat, strategic bombing—both daylight and night—the use of naval carriers to project airpower, and the first and only uses of atomic bombs during hostilities. The following case study asserts that we also saw the first widespread and well-orchestrated use of information warfare, and presents a hypothetical model for interaction between deception and cryptanalysis.

Many of us remain intrigued by the clandestine and covert operations conducted by the Allies in WWII. This study discusses two of those operations: deception and cryptanalysis based on radio intercepts. It also, and more importantly, attempts to build a model for an interactive relationship between the two that could have synergistically improved the contributions of these operations to the successful prosecution of the war. The model, though purely hypothetical, uses facts to present a case for the potential of maximizing misinformation through the integration of these two disciplines. Said another way, this paper suggests that the Allied leadership, specifically Winston Churchill, found cryptanalysis necessary but not sufficient for victory. Cryptanalysis and deception were both necessary and sufficient. Hence, the logic of the model suggests that Churchill directed an offensive information warfare campaign.

The Logic of the Model

The question posed is whether Prime Minister Churchill would, or could, have selectively chosen to chance using the Enigma (the machine used by the Germans to encipher high-grade wireless traffic[10]) to encipher notional messages and intrude on German wireless radio nets to misinform the Germans on Allied intentions or otherwise disrupt German military operations.

Churchill's concern for the security of the Enigma, and the knowledge that it was being used by the Allies, was considerable, as will be shown later. The risk of compromising Allied use of the Enigma was colossal, affecting many lives and the potential outcome of many battles. On the other hand, successful deception could be equally effective.

For Churchill to have taken this step would have been boldness from "sheer necessity" in the strictest Clausewitian terms.[11] The risk in not doing so would have to have been greater than the risk in doing so. The logic of the model is that if Churchill directed that messages encrypted using Enigma be transmitted, he would only have done so out of necessity—when Britain was in dire straits.

"Deception is as old as war itself."[12] Although this statement is from WWII, it was clearly not a revelation of fact. Sun Tzu included deception as one of his tenets of warfare when he said, "All warfare is based on deception."[13] The modern complexities of war and the ensuing technological advancements enhance the means through which deception can be employed, and WWII was no exception. The use of deception during WWII has been widely publicized. At least one book, *The Man Who Never Was*, was published and a movie by the same title was made about a single event.[14]

Deception and its implementation occur in both the strategic and tactical spheres. The example documented above was strategic in its support of the Normandy invasion. Tactical deception at that time was thought to fall under three headings, visual, aural (or sonic), and radio.[15] While aural deception might apply in limited fashion to specific engagements, it is logical that visual and radio deception could be used for broader objectives at both the strategic and operational levels.

It is not surprising that most information concerning deception activities remained classified for many years after the war and is only now coming to the attention of the public. It appears that most, if not all, of the information concerning tactical deception has been declassified. This is not the case with another stratagem used against the Germans, that of intercepting radio communications and using the Enigma machine to decipher the message transmissions. While previously classified documents concerning Ultra are now largely available to the public, a review of the primary sources reveals that many still contain blank pages that are marked "not releasable" while others contain portions that have been blanked out with no explanation. Thus, even though we know much more today than we did 15 years ago about these activities, public access remains unavailable for much of it.

These continuing restrictions may well be the result of comments made on 15 April 1943 by Col Alfred McCormack in a memorandum to Col Carter W. Clarke. McCormack, then "Mr McCormack," had earlier been appointed as special assistant to the secretary of war to study the uses of Ultra and establish procedures for making the best use of this source. At the time of the memorandum, McCormack was deputy chief of the Special Branch and worked for its chief, Colonel Clarke. The purpose of the Special Branch was to handle signals intelligence. McCormack's memorandum consists of 54 pages on the origin, functions, and problems of the Special Branch, Military Intelligence Service (MIS). In this memorandum, McCormack describes, in his view, Ultra security requirements as follows:

> One lapse of security is all that is necessary to dry up a radio intercept source. Therefore, both on the officer level and below, only persons of the greatest good sense and discretion should be employed on this work. This consideration is basic since intercept information involves a different kind of secrecy than does most other classified information. It will make no difference a year from now how much the enemy knows about our present troop dispositions, about the whereabouts of our naval forces or about other similar facts that now are clearly guarded secrets. But it will make a lot of difference one year from now—and possibly many years from

> now—whether the enemy has learned that in April 1942 we were reading his most secret codes. Not present secrecy, not merely secrecy until the battle is over, but permanent secrecy of this operation is what we should strive for.[16]

This secrecy was maintained throughout the war. Only carefully selected individuals in Washington and in the field had access to the information produced through these intercepts. The procedures for use by field commanders and their personnel, including controls established to protect the information and its source were laid out in a letter to General Eisenhower from General Marshall on 15 March 1944.[17] These procedures lasted at least through the end of the war.

The Origin of Ultra

Ultra's origin begins with the delivery of a German Enigma machine to the British by Polish dissidents. The history and acquisition of the Enigma machine are quite lengthy and complex. It is sufficient here to reflect that the Poles had established a successful cryptanalytic effort against the Germans by the early 1930s, having begun their efforts in the early 1920s.[18] Using their own copy of the Enigma, they achieved their first successful break in reading Enigma ciphers in December 1932 and January 1933.[19] Between 1933 and 1939, successful reading of Enigma traffic was purely a Polish achievement.[20] Once the Enigma fell into British hands, however, they took the lead and used it successfully throughout the war.

Enter Winston Churchill

Winston Churchill had a profound interest in the Ultra traffic produced from Enigma and required that all important decrypts be provided to him.[21] His interest in codebreaking is documented as early as November 1924 when, as the chancellor of the exchequer, he requested access to intercepts.[22] In his request, he stated, "I have studied this information over a long period and more attentively than probably any other Minister has done. . . .I attach more importance to them as a means of forming a true judgment of public policy in these spheres than to any other source of knowledge at the disposal of the State."[23]

Then in September 1940, after only four months as the prime minister, he directed he be provided "daily all Enigma messages."[24] When this traffic became overwhelming in volume, he backed off to receiving several dozen messages a day.[25] During a visit to Bletchley Park, the headquarters for the British cryptanalytic organization, he spoke to a crowd of the station managers and referred to them as "the geese that laid the golden eggs and never cackled."[26] After the war, Churchill reiterated his faith in Ultra, describing it as his "secret weapon"[27] and stating his belief that "it had saved England."[28]

Churchill's concern for security of Ultra was paramount. He directed that no action be taken in response to Ultra intercepts unless cover could be provided,[29] and he, in fact, repeatedly allowed naval convoys to come under U-boat attack rather than risk compromising Ultra security.[30]

Churchill was also directly involved in the conduct of deception operations. He established the London Controlling Section (LCS) in his headquarters specifically to plan those stratagems necessary "to deceive Hitler and the German General Staff about Allied operations in the war against the Third Reich."[31]

Not only did Churchill establish the LCS but he also personally conceived the idea for this organization after a series of successful uses of deception in the Libyan desert led to the defeat of Italian forces. In one of those instances, a small British force of 36,000 men defeated an Italian force of 310,000 using deceptive measures. Realizing he was outnumbered and about to be overrun, the British commander used inflatable rubber tanks, field guns, two-ton trucks, and prime movers to present the image of a larger force. He employed crowds of Arabs with camels and horses to drag harrow-like equipment to stir up dust storms, and he used antiaircraft artillery to keep the Italian reconnaissance aircraft high—precluding them from sorting out the actual order of battle on the ground.

The Italians perceived a force on their right flank much larger than theirs and tried to run. Using only two divisions, the British captured 130,000 prisoners, 400 tanks, and 1,290 guns. Their losses were minimal for the magnitude of the conflict—500 killed in action, 1,400 wounded, and 55 missing in action.[32] This impressive event rocked London and gave credence to further development of this capability. This action, and others similar to it, convinced Churchill that deception needed an institution so it could be applied on a broader scale. Thus, LCS was born.

The LCS was the first bureaucracy designed expressly to deceive.[33] It was "members of the LCS and those of other British and American secret bureaus"[34] who developed and executed LCS activities, referring to their weapons as "special means."[35] In this context, "special means" is "a vaguely sinister term that included a wide variety of surreptitious, sometimes murderous, always intricate operations of covert warfare designed to cloak overt military operations in secrecy and to mystify Hitler about the real intentions of the allies."[36]

Back to Ultra—Its Contribution

Ultra proved its value as early as mid-July 1940 when it provided forewarning of German plans to attack England. Intercepts at that time revealed Hitler's directive outlining the planned invasion of England. The invasion was to begin with an air campaign. These intercepts continued, reaching a point of two-to three-hundred per day—all being read at Bletchley Park. On 13 August, when the first air raids began, the British were more informed of the plans than were many of the Luftwaffe units.[37]

Clearly, Ultra intercepts provided the bulk of intelligence to the Allies during the war. By June 1944, 90 percent of the European intelligence summaries provided to Washington were based on Ultra information.[38] Ultra provided information on force disposition and German intentions at both the strategic and tactical levels. Ralph Bennett describes Ultra's contribution succinctly in his preface to *Ultra in the West*:

> For by often revealing the enemy's plans to them before they decided their own, Ultra gave the Allied Commander an unprecedented advantage in battle: since Ultra was derived from decodes of the Wehrmacht's wireless communications, there could be no doubt about its authenticity, and action based upon it could be taken with the greatest confidence. So prolific was the source that at many points the Ultra account of the campaign is almost indistinguishable from the "total" account.[39]

Ultra information also has been described as "more precise, more trustworthy, more voluminous, more continuous, longer lasting, and available faster, at a higher level, and from more commands than any other form of intelligence."[40] It even provided information on German intercepts and analysis of British and American radio networks.[41] Taking advantage of this latter knowledge, the Allies established an elaborate communications network designed expressly to transmit bogus traffic that would misinform the Germans of their intentions and operations. What would have prevented including encrypted Enigma messages directly to the Germans in this bogus traffic?

Radio Deception

The British and Americans used manipulation through cover and deception to target specific sources of enemy information. For example, they released false information to the world press and staged activities that "made the news." They deceived enemy air reconnaissance through the maneuver of real troops, use of controlled camouflage (both to conceal and intentionally show indiscretions), dummy equipment, and "Q" lighting (the positioning of lights to draw bombers to nonexisting airfields).

Aware that German radio intercept units were targeting their transmissions, they used a three-pronged strategy against the German listening stations. First, they prepared notional radio traffic to be transmitted by special deception troops over nets established solely for the purpose of deception. Second, they sent notional radio traffic over authentic operational nets. Finally, they regulated the genuine traffic passed on authentic operational nets, creating dead time and peak traffic levels.[42] Signal troops employed in deception activities were specially trained in these operations[43] and thoroughly indoctrinated on the sensitivities that accompanied their efforts. The following statement was among the many instructions concerning security provided to them:

> You must realize that the enemy is probably listening to every message you pass on the air and is well aware that there is a possibility that he is being bluffed. It is therefore vitally important that your security is perfect; one careless mistake may disclose the whole plan.[44]

One of the most elaborate schemes employing radio deception was used in support of the First US Army Group (FUSAG), a notional, fictive organization headed by Gen George S. Patton,

Jr.[45] Conceived as a part of Bodyguard,[46] the FUSAG was composed of more than fifty "divisions" located in southeast England. Aware that the Germans anticipated an Allied attack, the purpose in establishing a nonexistent FUSAG was to persuade the Germans that the attack would take place at Pas de Calais.[47]

The radio net supporting FUSAG represented the following units: a Canadian army, a US army, a Canadian corps, three US corps, a Canadian infantry division, a Canadian armored division, six US infantry divisions, and four US armored divisions.[48]

The Case

Enigma traffic provided the tip-offs to the planned German invasion of Britain well in advance. The speed with which Bletchley Park was reading the German Enigma permitted the British cryptanalysts to extract intelligence from several hundred messages a day, even though the Enigma settings were complex and changed frequently.

The Enigma used wheels that had to be set in the proper order for the decryption to take place. These settings were usually changed every 24 hours with minor settings changed more often. Other minor settings were made with each message. The tip-off to the receiver for these latter settings was contained in the transmission.[49] The speed with which these messages were deciphered could have provided the essential information required by the British to use the machine to other advantages.

From the volume of intercept, it is obvious the British knew their targets' organizations and frequencies. The traffic would have provided them with information on message originators, addressees, associated organizations, and formats—allowing them to reconstruct necessary elements of the German radio communications network. The German use of "standard phrases, double encipherment. . . their lack of an effective, protective monitoring program, and their unshakable—even arrogant—confidence in Enigma"[50] made it unlikely they would use authentication devices in their messages. The Germans then clearly were vulnerable to deception efforts using encrypted Enigma messages broadcast by the nets serving the London Controlling Section. Would Churchill have taken the risks associated with exploiting this vulnerability?

Dire Straits

The Battle of Britain and the Normandy invasion were two of the most significant events in WWII. The Battle of Britain, particularly, represented a critical period for the British. The defeat of Germany in that battle required the all-out effort by Britain. The battle began with each side roughly equivalent in front-line fighters, but it was touch and go until the Luftwaffe lost its ability to mount sustained attacks.[51] The dangers facing the British during the massive air raids might have convinced Churchill at some point that it would be worth the risk to use the Enigma to intrude on German radio nets. Perhaps relying on the confusion and disorder he knew existed among some of the Luftwaffe units,[52] his assessment as to the potential for success could have led to this risky decision.

From 13 August until mid-September, 1940, the Luftwaffe conducted raids during daylight hours, and Ultra traffic revealed most, if not all, the targets that were to be hit. Interestingly, beginning in mid-September and lasting throughout October, the raids were flown at night, and the only target references available through Ultra were code names representing target locations. Had something tipped the Germans their mail was being read?

On 14 November, Ultra revealed Coventry as a target and at least one British official believed naming the town instead of using a code word was a mistake on the part of the Germans.[53] The use of code words surely made Churchill nervous, giving him cause to question if British use of Enigma had been compromised. This concern could account for his widely reported decision to take no action to evacuate Coventry other than to alert fire, ambulance, and police units.[54]

The Normandy Invasion was the last critical juncture for the Allies. A successful invasion would bring Germany and the Third Reich to their downfall. In preparing for the invasion, Operation Bodyguard had already been implemented.

The infrastructure for radio deception was in place and in use. This infrastructure would also have made an excellent point of origin for intrusion into German radio nets, using the Enigma to encipher messages for transmission. Schemes could have been devised using notional traffic sent over the deception nets, which were known to be monitored by the Germans, to complement intrusion traffic enciphered with Enigma. Bletchley personnel could prepare the Enigma traffic and send it to the radio deception units to be transmitted verbatim on specified frequencies. The personnel employed in the radio deception were well trained for their purpose and indoctrinated in the secrecy of their work.

If Churchill saw the invasion as the last big push to defeat Germany, he may also have viewed selective use of intrusion as justified and worth the risk. Given the increasing disruption that occurs with the multiplying intensity of battle, the risk would have gradually diminished with time during the course of the fight. As the risk diminished, the opportunities would have grown. Greater opportunities would have been enticing to Churchill, especially if there were opportunities to shape the postwar world.

Much of the history of WWII may need to be rewritten because of the revelations of Ultra contributions. Revelations include those already made and those yet to be made. Considering what we now know about Ultra operations, one can assume that credit for success in a battle often went to the wrong party. The men and women at Bletchely Park and other locations, who were involved in providing advance warning and other information to Allied forces may never get all the credit they are due. It is now well known that "Ultra did indeed shape the character of strategy and operations—particularly operations. In no other war have commanding generals had the quality and extent of intelligence provided by Ultra."[55]

Whether Churchill actually used the Enigma offensively for the purposes hypothesized here may never be known. If he did not, maybe the cause was that it was too risky, or just too tough to do. Maybe the Allies did not possess enough information on the keying cycles necessary to exploit that avenue of deception. Or maybe the Allies just missed a good opportunity. Absent further declassification, we cannot know for certain. While logic suggests Churchill would have exploited Enigma, the facts may prove otherwise.

If he did use it in this manner, it would have been information warfare at its best. Perhaps it was.

Impact of Information Technology on the Gulf War

Although the use and exchange of information have been critical elements of war since its inception, the Gulf War was the stage for the most comprehensive use of information, and information denial, to date. New technologies in this conflict enhanced the Coalition's ability to exchange and use information and highlighted the imperative of denying the adversary his ability to communicate with his forces.

While in large part these technologies were space-dependent, recent advancements in digital technology permitted the rapid processing, transmission, and display of information at all echelons, enabling decision makers to respond rapidly to developing situations on the battlefield. Some prototype systems, such as JSTARS, successfully made their trial run during this conflict, earning their place in history as contributors to the Coalition success in this war.

Architectures enabling connectivity between these many systems were nonexistent when Iraq invaded Kuwait; however, they were put in place during the buildup and supported Coalition forces for the duration of the war. These architectures were clearly necessary to effectively control the myriad activities operating simultaneously in the battlefield.

For example, 11 Airborne Warning and Control System (AWACS) aircraft controlled 2,240 sorties a day, more than 90,000 during the war, with no midair collisions and no friendly air engagements. Satellite connectivity permitted this same air activity to be displayed live in the Pentagon command center.

JSTARS tracked tanks, trucks, fixed installations, and other equipment, even though this system had not met operational capability status. Satellites, microwave, and landlines handled 700,000 phone calls and 152,000 messages a day. Coalition forces avoided communications interference through successful deconfliction of more than 35,000 frequencies. Any attempt to describe the complexities of managing this system would be an understatement.

The Joint Communications-Electronic Operating Instructions (JCEOI), which was used to allocate frequencies, call signs, call words, and suffixes for the Gulf War, was published in over a dozen copies and weighed 85 tons in paper form.[56] This system was used for both space and terrestrial communications.

Gulf War Space Contributions

Space assets, both military and commercial, belonging to the United States, the United Kingdom, France, and the USSR provided the Coalition with communications, navigation, surveillance, intelligence, and early warning, as well as offering live television of the war to home viewers around the world for the first time.

Using some 60 satellites, Coalition forces had access to secure strategic and tactical communications in-theater and into and out of the theater of operations.[57] These satellites bridged the gap for tactical UHF and VHF signals that heretofore had been limited to terrestrial line of sight. Thus time-sensitive information could be exchanged between ground, naval, and air units spread throughout the theater. Without this capability, the communications required to support the preparation and distribution of task orders and the coordinated operations of AWACS, JSTARS, and conventional intelligence collection in support of force packages in virtual and near-real-time would have been impossible. Even though there were still shortfalls at the tactical level in timeliness, precision, and volume, commanders at all levels had access to unprecedented communications capabilities.

There are some who credit the capabilities afforded by the NAVSTAR Global Positioning System (GPS) "as making the single most important contribution to the success of the conflict."[58] Using a constellation of 14 satellites, Coalition forces were able to locate and designate targets with remarkable precision, navigate through the naked Iraqi desert better than the Iraqis themselves, and find troops in distress faster than ever before. The US Army used the GPS to navigate the Iraqi desert in the middle of sand storms, surprising even the Iraqis, who themselves do not venture across it for fear of becoming lost.

GPS capability made possible the "left hook" used to defeat Saddam Hussein's armored divisions.

The use of GPS was, in large part, the result of off-the-shelf purchases acquired by special contract arrangement; these were the same systems that had been designed and marketed for recreational boat use—thus, technically available to anyone.[59] US troops stationed in Saudi Arabia also received commercially purchased GPS devices from their relatives.[60] Access to GPS, and its attendant capabilities, added tremendously to the morale of Coalition forces.

More than 30 military and commercial surveillance satellites were used for intelligence gathering during the war.[61] These satellites provided Coalition forces with imagery, electronic intelligence, and weather data. While these systems provided precise targeting information on enemy locations, movement, and capabilities, they were also essential in meeting another Coalition objective—that of minimizing collateral damage. Precision targeting combined with the use of precision guided munitions significantly decreased civilian casualties and left structures adjacent to targets intact.

Gulf War Intelligence

The rapid deployment of a variety of systems to the Persian Gulf in response to the crisis there led to a number of stovepiped organizations, resulting in a voluminous amount of unfused and uncorrelated information being collected and disseminated. Also many incompatible systems were deployed. This lack of integrated, all-source information and the deficiency in compatibility often placed a burden on recipients who had neither the personnel nor the skills necessary to put it all together in one product.

Notwithstanding this limitation, the bulk of which involved secondary imagery production, the evidence shows that timely, quality intelligence was available to those units fortunate enough to have access to the right terminal systems. To a large degree, the impediment was the result of fielding prototype systems for which there was little terminal capability.

One of the most prolific producers of information in this category was the Tactical Information Broadcast Service (TIBS), but a limited number of terminals dictated that only key nodes could have access to this product. Nevertheless, TIBS and its cousin, Constant Source, provided timely updates of intelligence information to various echelons, including wings and squadrons, directly from collectors and associated ground processing facilities.[62]

The RC-135 Rivet Joint, flying in coordination with its sister ships, the E-3 AWACS and E-8 JSTARS, flew 24 hours a day to support the war. Referred to as the "ears of the storm" in contrast to the AWACS role as "eyes of the storm,"[63] the RC-135 provided real-time intelligence to theater and tactical commanders in the desert and Persian Gulf areas. Specially trained personnel used on-board sensors to identify, locate, and report Iraqi emitters that might pose a threat to Coalition forces.

These systems are only a sampling of those deployed to the theater to provide intelligence support. Reviews and action are ongoing to resolve the problems resulting from stovepiping and incompatible systems.

Iraqi Command and Control (Or Lack Thereof)

The Coalition not only recognized the value of information to its efforts, it also saw the benefits of denying the Iraqi command and control system its ability to function. The Coalition identified the Iraqi leadership and Iraqi command, control, and communications (C^3) facilities as the key centers of gravity.[64] While command of the air was the initial key objective, C^3 facilities received priority in targeting.

The Coalition used massive airpower at the onset of hostilities to accomplish this objective. Targeting strategic military, leadership, and infrastructure facilities, the Coalition launched its attack on Iraq on 17 January 1991. Early warning sites, airfields, integrated air defense nodes, communications facilities, known Scud sites, nuclear/chemical/biological facilities, and electrical power facilities were attacked by B-52s, Tomahawk land-attack missiles (TLAMs), F-117s, and helicopter gunships. During the first two days, the Coalition gave no slack while conducting the most comprehensive air attack of the war.

After only the opening minutes of the war, Iraq had little C^3 infrastructure remaining.[65] The Coalition success was so devastating that, as an Iraqi prisoner reported, "Iraqi intelligence officers were using Radio Saudi Arabia, Radio Monte Carlo, and the Voice of America as sources to brief commanders."[66] What little communications capability Iraqi tactical commanders did have, they used improperly.

Apparently concerned over Coalition communications monitoring, the Iraqis practiced strict communications security through near total emission control (EMCON). While this did have a negative effect on Coalition signals collection efforts, it also blinded Iraqi tactical units. One Iraqi brigade commander, in reflecting his surprise over the speed with which a US Marine unit overran his unit in Kuwait, showed he had no idea the Marines were coming even though another Iraqi unit located adjacent to him had come under attack two hours before.[67]

Although leadership as a target was difficult to locate and survived the conflict, the successful attacks against Iraqi C^3 essentially put her leadership in the position of having no strings to pull. Trained to operate under centralized control, Iraqi forces did not know how to function autonomously. Air defense forces became fearful of emitting because of their vulnerability to antiradiation missiles. Believing the army, not the air force, was the determining force in battle, the Iraqis attempted to shield rather than use their aircraft. The attempts they did make in defensive counterair proved rather embarrassing.

Gulf War Conclusions

The Gulf War clearly demonstrated the need for accurate and timely dissemination of information. Information was the hub of all activity on the Coalition side, and the lack of it caused the failure of the Iraqi military to employ its force. The communications enhancements realized with the advent of new technologies also brought about new vulnerabilities.

Building defenses to these vulnerabilities is considered by some to be at odds with increasing the capabilities. The benefits enjoyed by the Coalition's ability to communicate and the impact of attacks on Iraqi C^3 have been widely publicized and have to be assumed to be well known by every potential adversary. We have to prepare for similar attacks, or attacks of a different medium, against our own information systems in the future.

What Does the Future Hold?

There is an information glut. There is a proliferation of modem-equipped personal computers and local area networks in military organizations, industrial facilities, and private homes around the globe. And it does not stop there. For example, Motorola is working on a 77-satellite constellation that will provide cellular telephone service from any spot on earth within five years. With fiber optics supporting these satellites, entire countries are being wired. Turkey, for example, has moved into the information age in one big leap.[68]

As the information glut continues to grow, along with systems to accommodate it, vulnerabilities to surreptitious entry are certain to increase. The amount of information being reported is doubling every 18 months. And this growth is accelerating. Two years ago, volume was doubling every four years; three years ago, it took four and a half years.[69] While our capacity to process information at this growth rate seems limited, technology has a way of catching up—but not necessarily in time to help for a given situation. It can be particularly difficult to process large amounts of information in a readily useable form during intense, crisis situations.

During the Gulf War, 7,000 personnel worked two days to produce the air tasking order (ATO) for 2,000 aircraft sorties to be flown on the third day. The ATO began as a 300-page document developed for transmission to Air Force, Navy, and Marine aviation, but difficulties at receiving organizations forced adjustments.

Even using dedicated communications circuits, it took the Navy three to four hours to receive the ATO. Early on, there was a 70,000-message backlog, and flash-precedence messages were taking four to five days to reach their destination—some never made it. Additionally, the volume of traffic took an inordinate amount of time to read, let alone respond to.[70] It seems the greater our capability to process information, the more information there is to process.

Former vice-chairman of the JCS, Navy Adm David E. Jeremiah, sees it this way: "Technology has fueled a change in communication, [ushering in] an era of information dominance. Global dominance will be achieved by those that most clearly understand the role of information and the power of knowledge that flows from it."[71]

The services are recognizing this change in communications, and reacting to it. In the Air Force, information warfare techniques are being intensively studied and incorporated at the Air Intelligence Agency (AIA). AIA looks at information dominance in terms of the Observe, Orient, Decide, Act (OODA) loop. The OODA loop represents the decision cycle through which a warrior at any level must go. As you go from the strategic level to the tactical level, the time available for making a decision decreases. At the tip of the spear, it is very short.

According to General Minihan, "As we compare friendly and adversary OODA loops, it becomes a deadly game of compression and expansion. We will use information warfare to expand the adversary's and compress our own action loops. If you can't think, can't hear, and can't see—and I can—you will lose every time."[72] This concentration of effort in information technology will, and should, have an impact on military doctrine.

Admiral Jeremiah has already considered this. He points out that "it is time to come to grips with a different intersection, an intersection of technology and strategic thought. . . . I think that in large measure the product today, technology, drives doctrine and tactics, and to a major degree drives strategy."[73]

We obviously are far from reaching full understanding of the impact of information warfare on doctrine, tactics, and strategy. However, the explosion of information on societies around the world, and the associated technology, dictate that we find a way to measure the impact, and look for ways to incorporate the right level of emphasis on this topic into our thinking. One area of concern is our propensity to stovepipe activities within our structures, and the negative influences this can have on military operations.

Army, Navy, and Air Force senior leaders have voiced concern with these vertical structures. It has become tradition, for example, to stovepipe several functional areas such as intelligence, logistics, and acquisition. Stovepiping often excludes the chain of command from the decision-making process and impedes synergistic benefits that are available from integrated operations.

The focus, then, should be on moving from vertical structures, or stovepipes, to horizontally integrated systems. The expected result is integrated functional areas, which should provide a better structure for identifying needs and requirements, and determining force projection priorities. In the information sphere, however, this could increase vulnerabilities to unauthorized access because it disperses the information base on a much wider scale. Some members of the US military community recognize that "interdicting, protecting, and exploiting these new pathways is what IW (information warfare) is all about."[74] As we place more emphasis on this new dimension, we can expect other nations to follow. Russia will probably be one of the first.

Russian senior military officials have already recognized that the integration of information technology "could generate radical changes in the organizational principles of armed forces."[75] The use of "intellectualized" weapons in the Gulf War by the Coalition apparently sparked a move in the same direction in Russia. Russian military experts now believe in "a new axiom to the body of military art: For combatants contending in military conflict today, 'superiority in computers' is of precisely the same significance as superiority in tube artillery and tanks was to belligerents in earlier wars."[76]

Furthermore, "superiority in the MTR [military-technical revolution] proceeds from superiority in 'information weapons': 1) reconnaissance, surveillance, and target acquisition systems, and 2) 'intelligent' command-and-control systems."[77] Russian military leaders believe the new "formula for success" is to "First gain superiority on the air waves, then in the air, and only then by troop operations."[78] As the two former adversarial world superpowers, who by and large supplied most of the weapons to other countries around the world, pursue information warfare as a new realm of combat, it is almost certain other nations will buy into the trend.

In what is probably only the beginning for nations in conflict, the Internet has already provided a medium for information warfare between two belligerent nations. During the recent border dispute between Ecuador and Peru, Ecuador used the Internet to publish government bulletins and excerpts from local media to tell its side of the conflict. In retaliation, Peru Internet used a gopher site in an attempt to neutralize Ecuadorian propaganda. [A gopher is an information system residing on the Internet that knows where everything is and, through an arrangement of nested menus, allows a user to continue choosing menu items until the sought-after subject is located.[79]] The resulting verbal skirmish left both nations working to set up their own gophers.[80]

Global information systems will enable ordinary users to access an extraordinary number of databases, far beyond the Internet capability of today (which is more than a million files at databases located at universities and corporate research centers). New software technologies permit these accesses to be conducted autonomously, using "self-navigating data drones."

These drones, referred to as "knowbots," are released into the Internet and search for information on their own. They can roam from network to network, clone themselves, transmit data back to their origin, and communicate with other knowbots.[81] Given this capability, one has to wonder, and perhaps be concerned, about the potential for unauthorized, or at least undesirable, access to certain databases and computer activities.

Hackers routinely attempt to get into US military systems. During the Gulf War, hackers from Denmark, Moscow, and Iraq tried to penetrate these systems.[82] Our awareness of these attempts does not necessarily prove there were no successes of which we are unaware. And, even if they failed during that conflict, can we guarantee the security of our systems during the next war?

These vulnerabilities were revealed recently when a British teenager using a personal computer at his home hacked his way into a US military computer network, gained access to files containing sensitive communications relating to the dispute with North Korea over international inspections of its nuclear program, and, after reading them, placed them on the Internet. His actions made those files available to about 35 million people. Officials suspect he had access to these computers for weeks, perhaps even months, before he was caught.

Interestingly, once it was known an intruder was in the system it only took a week to identify him. Unfortunately, the apparent difficulty was in detecting him. Officials added that he had also breached other defense systems.[83]

Paul Evancoe and Mark Bentley, computer virus experts, have documented their concerns over our vulnerability to computer virus warfare (CVW) by other nations. They describe in detail the vulnerability of computer systems to this danger, and claim that "CVW is a powerful stand-alone member of the non-lethal disabling technology family and is likely being developed by several countries."[84]

They also point out that the intelligence community and policy makers do not focus on these threats and generally do not possess enough technical understanding to recognize CVW as a real national security threat. They believe CVW remains an abstract, nontangible concept to most intelligence analysts and policymakers. Furthermore, they call for legislation outlawing CVW development, classifying CVW as a weapon internationally, and including it as part of nonproliferation treaties.

It is unrealistic to believe we could achieve the support of the international community in this regard, and, with our lead in technology, we probably do not want to do so. Even if we could acquire this level of cooperation, and wanted to, enforcement would be next to impossible. CVW development does not leave traces as does chemical, biological, and nuclear development. And our efforts to isolate those are not always met with success.

Some Americans believe there will be no big wars in the future because there is too much destructive power, and nobody wins. The interdependence of nations would likely result in as much damage to an aggressor as to its adversary. Whether this is true or not, the concept of national security is changing.[85] Among the threats we face today are terrorism—either state-sponsored or radical element, proliferation of weapons of mass destruction, localized conflicts, and aggressors that upset the world peace balance, intense economic competition, and availability of food and water.

The US military may be called upon to react, in one way or another, to any of these threats. US military operations can run the gamut, from civil-military affairs assistance to forcible entry. More reliance is being placed on communications and intelligence systems in support of these activities, and as these systems become more interoperable, they may become more vulnerable. "It is becoming more and more difficult to distinguish C^4 systems from intelligence systems."[86] While sophisticated antijam systems are being developed and deployed, these systems are still computer based. Disruption in one would affect others.

For example, for years we have needed a near-real-time intelligence system capable of providing targetable accuracy information to "shooters." The Army expects to have an airborne and ground-based SIGINT/EW system capable of doing that by the end of the decade.[87]

It seems logical that other existing and developmental systems might also be interconnected. Some of these might include the Joint Targeting Network (JTN), the Tactical Information Broadcast Service (TIBS), Tactical Receive Equipment Related Applications (TRAP), Senior Ruby, Constant Source, Quick Look, Over-the-Horizon (OTH) systems, and air and ground-based radar systems. The integration and wide dispersal of these systems increase the number of vulnerability points where an adversary might intrude.

The GPS may be one of the most revolutionary systems in our inventory when you consider the difference it can make in navigation and geo-positioning of assets. It is available to the public and anyone with a few hundred dollars can buy into the system. The benefits, then, that we derive from this capability may be offset somewhat by use of the system by an adversary. GPS has improved our navigation and geo-positioning accuracy in multiples, but we are not the only ones who can use it.

Conclusions

Even though the anticipated national security threats of the coming decades involve less developed countries, the CVW threat and other methods of intrusion and disruption are not necessarily beyond their reach.

Opportunities to deceive and confuse through an elaborate misinformation scheme along a myriad of information paths are available to anyone. Information warfare provides a new avenue to employ deception techniques through the use of multiple paths that create the perception and validation of truth. These activities can put new light on Winston Churchill's statement at Tehran in November 1943 concerning Allied deception efforts, "In war-time, truth is so precious that she should always be attended by a bodyguard of lies."[88]

In this vein, General Minihan proposes the prospect of "an intelligence analyst manipulating an adversary's command and control system so that reality is distorted."[89] Consider Marvin Leibstone's projection, ". . . tomorrow's soldier will depend more than ever on the very well known and trusted factors of mobility and C^3I.[90] Imagine a scenario depicting a "left hook" in the Iraqi desert that fails because the systems in use were successfully attacked by CVW, or some other intrusion method, with the resulting disruption putting US troops in a flailing posture—facing the unknown and losing confidence in their operation. One thing is sure. An Iraqi "left hook" will be difficult to repeat. We have to assume Iraq, and others, will exploit the GPS to their own advantage. Information warfare is coming of age!

World War II set the stage, but only with today's technology can we expect action in this sphere of warfare on a grand scale. Fortunately, the US military senior leadership is becoming involved, and, in many cases, taking the lead on this perplexing issue. With this emphasis, we must carefully assess the vulnerabilities of the systems we employ. Systems proposals must be thoroughly evaluated and prioritized by highest value payoff. This needs to be accomplished through a more balanced investment strategy by the US military that conquers our institutional prejudices that favor "killer systems" weapons.[91] Offensive systems will be at risk if we do not apply sufficient defensive considerations in this process.

"The electromagnetic spectrum will be our 'Achilles heel' if we do not pay sufficient attention to protecting our use of the spectrum and at the same time recognize that we must take away the enemy's ability to see us and to control his forces."[92] We must also interdict the opportunities for adversaries to intrude on our systems. Other nations have realized the value of offensive applications of information warfare; therefore, we must attack the issue from two directions, offensively and defensively, with almost equal accentuation.

Information warfare adds a fourth dimension of warfare to those of air, land, and sea. When the Soviets developed a nuclear program after World War II, the United States was caught by surprise. In this new dimension, we must stay ahead.

Notes

1. Carl von Clausewitz, *On War*, ed. and trans. Michael Howard and Peter Paret (Princeton, N.J.: Princeton University Press, 1984), 84.
2. Sun Tzu, *The Art of War*, trans. Samuel B. Griffith (New York; Oxford University Press, 1971), 84.
3. Information Warfare: Pouring the Foundation, Draft, USAF/XO, 19 December 1994, i.
4. Ibid., 3.
5. Alvin and Heidi Toffler, *War and Anti-War: Survival at the Dawn of the 21st Century* (Boston: Little, Brown, and Co., 1993), 140.
6. Ibid, 139.
7. Craig L. Johnson, "Information Warfare—Not a Paper War," *Journal of Electronic Defense* 17, no. 8, August 1994, 56.
8. Ibid.
9. John H. Petersen, "Info Wars," US Naval Institute Proceedings, 119, May 1993, 85.
10. Peter Calvocoressi, *Top Secret Ultra* (New York: Pantheon Books, 1980), 3.
11. Clausewitz, 191.
12. John Mendelsohn, ed., *Covert Warfare: Intelligence, Counterintelligence, and Military Deception During the World War II Era* 18 (New York: Garland Publishing, Inc., 1989), 1. Note: This book is the last in a series of 18 volumes on covert warfare edited by Mendelsohn. The content of the series is primarily composed of declassified documents residing in the National Archives. These documents included classifications up through TOP SECRET ULTRA. The quoted material in this paper from this series is usually taken from the copied material of the original documents.
13. Sun Tzu, 66.
14. See Ewen Montagu, *The Man Who Never Was* (New York: J.B. Lippincott Co., 1954).
15. Mendelsohn, chap. 1, 1.
16. Ibid., vol. 1, *Ultra Magic and the Allies*, chap. 8, "Origins, Functions, and Problems of the Special Branch, MIS," 27.
17. Mendelsohn, vol. 1, chap. 4, "Synthesis of Experiences in the Use of ULTRA Intelligence by U.S. Army Field Commands in the European Theater of Operations," 4.
18. Wladyslaw Kozaczuk, *ENIGMA* (University Publications of America, Inc., 1984) chapter 2.
19. Ibid., 20–21.
20. Ibid., 95.

21. Ibid., 165.
22. David Kahn, *Seizing the ENIGMA: The Race to Break the German U-Boat Codes, 1939–1943* (Boston: Houghton Mifflin Co., 1991), 184.
23. Ibid.
24. Ibid.
25. Ibid.
26. Ibid.
27. James L. Gilbert and John P. Ginnegan, eds., *U.S. Army Signals Intelligence in World War II: A Documentary History, Center of Military History, United States Army* (Washington, D.C.: GPO, 1993), 175.
28. Ibid., 175.
29. Kahn, 276.
30. Gilbert and Ginnegan, 176.
31. Anthony Cave Brown, *Bodyguard of Lies* (New York: Harper & Row, 1975), 2.
32. Ibid., 50.
33. Ibid., 45.
34. Ibid., 2.
35. Ibid.
36. Ibid.
37. Kozaczuk, 156–66.
38. Mendelsohn, vol. 1, chap. 3, "Use of CX/MSS ULTRA by the U.S. War Department, 1943–1945," 17.
39. Ralph Bennett, *Ultra in the West: The Normandy Campaign 1944–45* (New York: Charles Scribner's Sons, 1979), viii.
40. Kahn, 276.
41. Bennett, 42.
42. Mendelsohn, vol. 15, *Basic Deception and the Normandy Invasion*, chap. 4, "Cover and Deception, Definition and Procedure, Exhibit '3' of C&D Report ETO," 1 and 2.
43. Mendelsohn, vol. 15, chap. 6, "Cover and Deception Recommended Organization, 8 September 1944, Exhibit '5' of C&D Report ETO," 2.
44. Ibid., vol. 15, chap. 10, "Operations in Support of Neptune: (B) FORTITUDE NORTH, 23 February 1944, Exhibit '6' of C&D Report ETO," Appendix 'C' to SHAEF/18216/1/Ops dated 10th March 1944.
45. Józef Garlinski, *The Enigma War* (New York: Charles Scribner's Sons, 1980), 159–160. After twice striking "shell-shocked" soldiers, Patton had gotten into trouble. General Eisenhower needed a place to put him, and, knowing the Germans kept track of his finest generals, considered Patton the perfect choice for this notional outfit. In Eisenhower's mind, placing Patton in charge would make this concoction more believable to the Germans.
46. Brown, 10. BODYGUARD was the cover name given to the deception plan developed for NEPTUNE, the cover term for the Normandy invasion. It was taken from Churchill's statement at Tehran, "In war time, truth is so precious that she should always be attended by a bodyguard of lies."
47. Garlinski, 160.
48. Mendlesohn, vol. 15, chap. 11, "Operations in Support of NEPTUNE: (C) FORTITUDE SOUTH I, Exhibit '6' of C&D Report ETO," appendix B, pt. I.
49. Bennett, 4.

50. Diane T. Putney, ed., *ULTRA and the Army Air Forces in World War II* (Washington, D.C.: Office of Air Force History, United States Air Force, 1987), 97.

51. H.P.Willmott *The Great Crusade—A New Complete History of the Second World War*, (New York: Free Press, 1989) 108–9.

52. Kozaczuk, 166.

53. Ibid., 167.

54. Ibid., 167.

55. Putney, 35. This statement was made by Associate Justice of the Supreme Court Lewis F. Powell, Jr., in an interview conducted by Dr Richard H. Kohn, chief, Office of Air Force History, and Dr. Diane T. Putney, chief, Air Force Intelligence Service Historical Research Office. During WWII, Justice Powell was one of a select group of people chosen to integrate Ultra information into other intelligence. As an intelligence officer in the Army Air Force, he served with the 319th Bomb Group, Twelfth Air Force, and the Northwest African Air Forces. He was on General Carl Spaatz's United States Strategic Air Forces staff as Chief of Operational Intelligence, as well as being General Spaatz's Ultra officer, towards the end of the war. He made at least one visit to Bletchely Park where he stayed and worked for several weeks.

56. David L. Jones and Richard C. Randt, "The Joint CEOI," in *The First Information War: The Story of Communications, Computers and Intelligence Systems in the Persian Gulf War* (Fairfax, Va.: AFCEA International Press, October 1992), 162.

57. Sir Peter Anson and Dennis Cummings, "The First Space War: The Contribution of Satellites to the Gulf War," in Alan D. Campen, ed., *The First Information War: The Story of Communications, Computers and Intelligence Systems in the Persian Gulf War* (Fairfax, Va.: AFCEA International Press, October, 1992), 121.

58. Ibid., 127.

59. Peterson, 85.

60. Anson and Cummings, 127.

61. Ibid., 130.

62. James R. Clapper, Jr., "Desert War: Crucible for Intelligence Systems," in Alan D. Campen, ed., *The First Information War: The Story of Communications, Computers and Intelligence Systems in the Persian Gulf War* (Fairfax, Va.: AFCEA International Press, October 1992) 82.

63. Robert S. Hopkins III, "Ears of the Storm," in Alan D. Campen, ed., *The First Information War: The Story of Communications, Computers and Intelligence Systems in the Persian Gulf War* (Fairfax, Va.: AFCEA International Press, October 1992) 65.

64. Thomas A. Keaney and Eliot A. Cohen, *Gulf War Air Power Survey Summary Report* (Washington, D.C.: Department of Defense, 1993), 40.

65. Alan D. Campen, "Iraqi Command and Control: The Information Differential," in Alan D. Campen, ed., *The First Information War: The Story of Communications, Computers and Intelligence Systems in the Persian Gulf War* (Fairfax, Va.: AFCEA International Press, October 1992), 171.

66. Ibid., 172.

67. Ibid., 174.

68. Petersen, 88.
69. Ibid., 89.
70. Ibid., 86.
71. John G. Roos, "InfoTech InfoPower," *Armed Forces Journal International*, June 1994, 31.
72. "Information Dominance Edges Toward New Conflict Frontier," *Signal International Journal* 48, no. 12, August 1994, 37.
73. Roos, 31.
74. Johnson, 55.
75. Mary C. FitzGerald, *The Impact of the Military-Technical Revolution on Russian Military Affairs*, vol. 2, Hudson Institute, submitted in partial fulfillment of Contract #MDA903-91-C-0190, HI-4209, 20 August 1993, 98.
76. Ibid., 100.
77. Ibid.
78. Ibid.
79. Tom Lichty, *The Official America Online for Windows Tour Guide*, 2d ed., ver. 2, 325.
80. *Newsweek*, 20 February 1995,12.
81. Peterson, 89.
82. *Army Times*, 54, no. 43, 23 May 1994, 28.
83. *Baltimore Sun*, 9 January 1995, 3.
84. Paul Evancoe and Mark Bentley, "CVW—Computer Virus as a Weapon," *Military Technology* 18, no. 5, May 1994, 40.
85. Petersen, 90.
86. Darryl Gehly, "Controlling the Battlefield," *Journal of Electronic Defense*, 6, no. 6, June 1993, 48.
87. Gen Jimmy D. Ross, "Winning the Information War," *Army*, February 1994, 32.
88. Brown, 10.
89. "Information Dominance Edges Toward New Conflict Frontier," *Signal*, 48, no. 12, August 1994, 39.
90. Marvin Leibstone, "Next-Generation Soldier: Ditched, or Digitized?," *Military Technology* 18 no. 7, July 1994, 59.
91. "Army Plan Fosters Dynamic Information War Framework," *Signal International Journal* 48, no. 3, November 1993, 56.
92. Ross, 28.

Overview: Biological Warfare Issues

Biological warfare (BW) is the use of disease to harm or kill an adversary's military forces, population, food or livestock. Living organic germs, like anthrax, are a major example of biological weapons. Another is byproducts of organisms, known as toxins. An example is botulism. Biological agents are much deadlier, pound for pound, than chemical agents. It has been estimated that 10 grams of anthrax could kill as many people as a ton of the nerve agent Sarin.

Lt Col Terry Mayer traces the history of biological warfare, including its use during World War II by Japanese forces in China. During early phases of the Cold War the United States, as a potential retaliatory measure against the USSR, developed a BW program. However, in 1969 President Nixon terminated the program and announced US unilateral disarmament of offensive BW weapons. A worldwide Biological and Toxin Weapons Convention (BWC) followed, and among the 118 signers are the United States, Russia, and Iraq. However, there is evidence that Moscow continued to manufacture anthrax at least into 1992. Iraq now admits that it, too, had a BW program until 1991.

The BW problem was dramatically highlighted during Desert Storm when Iraq was discovered not only to have a nuclear weapons program, but also an elaborate chemical and biological warfare effort. US government analysts are skeptical about whether or not Iraq actually destroyed its biological warfare agents and equipment in 1991 as they claim. Some believe that Baghdad may still have a large BW program intact. Unclassified information from the Defense Nuclear Agency documents indicate that numerous rogue states—those that support state sponsored terrorism like Iran, Iraq, Libya, and North Korea—have or are pursuing BW programs.

Colonel Mayer indicates that BW agents of other states are very difficult, if not impossible, for allied intelligence services to detect in research, production, transit, or employment phases. In addition to detection shortfalls, Mayer's essay questions whether the United States is unable to effectively protect its military forces (medically and nonmedically) from

BW weapons, conduct an effective preemptive counteroffensive strike against enemy BW facilities, or protect the civilian population against a terrorist BW attack.

Robert Kadlec, M.D., (Lt Col, USAF) shows that the ongoing revolution in biotechnology has also made possible medical products readily transferable to biological warfare applications. Some vaccines are available, but do not have wide distribution. He contends that the proliferation of BW weapons provides less-developed countries capabilities that could be as lethal and devastating as nuclear weapons. BW weapons are inexpensive, easy to produce, can be disguised as natural events (for example as agricultural sprayers), are hard to detect or preempt, and are hard to defend against once employed.

Dr Kadlec analyzes the current BW threat and cites Office of Technology Assessment (OTA) findings which emphasize how certain unstable states have governments seeking WMD, including BW. The OTA analysis suggests that BW is the cheapest and most easily produced of all the weapons of mass destruction. Indeed some third world regimes see BW as a "potential equalizer" to offset Western conventional or nuclear forces.

In his essay on biological warfare as a possible means of attacking an adversary's agricultural base, Dr Kadlec shows how "the existence of natural occurring or endemic agricultural pests or diseases and outbreaks permits an adversary to use BW with plausible denial." Dr Kadlec shows instances of how BW could be used in attacks on livestock and plants including anthrax, glanders, rinderpest, and wheat rust.

He notes instances of naturally occurring infestations of agriculture. For example, the whitefly infestations of California crops in 1981 and 1991 caused $500 million worth of damage. The Russian wheat aphid cost the United States $600 million. Similarly, in other cases the Mediterranean fruit fly caused $900 million in damage and lost revenues to American crops. To guard against such pests, over $7 billion is spent every year on pesticides.

Dr Kadlec then provides a series of illustrative and hypothetical scenarios of how biological warfare could be waged against certain food suppliers: spraying a corn seed blight over the Midwestern United States from commercial airliners to force massive corn imports; deliberately spreading a rapidly breeding grape louse across California wine country; deliberately sabotaging Pakistan's cotton crop with insects; etc. Thus, while not lethal attacks on human beings, such BW attacks against a country's food supplies or crops could be very damaging. Such BW attacks against a competitor's economic assets could open an unfortunate new chapter in the history of economic warfare.

Chapter 8

The Biological Weapon: A Poor Nation's Weapon of Mass Destruction

Lt Col Terry N. Mayer, USAF

Prologue—4 January

This is a CNN special report. This just in. The Center for Disease Control has just declared that an epidemic is widespread in Miami, Florida. Doctors have not yet diagnosed the specific cause of the rampant disease, but the illness initially resembles a chest cold that progresses into pneumonia-like symptoms. It then progresses rapidly into fever and shortness of breath. What is especially peculiar about this epidemic is that all the patients who have sought medical attention attended the Orange Bowl football game on New Year's Day. Authorities have asked that anyone who went to that game seek medical care if cold-like symptoms appear. Stay tuned to CNN for further developments on this story. Elsewhere in the news ...

The Biological Warfare Context

This is a notional, yet frightening illustration of what the first signs of a biological warfare (BW) attack might sound like. This scenario is a plausible example of an attack a terrorist or deranged person might conduct using off-the-shelf technology and readily available biological warfare agents. The "concept of operations" in this attack consisted of using several insect bombs (the kind where you push the button, it starts spraying, and you leave the house for two hours) and modifying them by filling the canister with anthrax bacteria bought through a mail order specimen company in the United States. If that doesn't sound credible, please note that Saddam Hussein bought his original anthrax culture from a mail order house in the United States and had it shipped overnight mail![1] This is just a sample of many plausible scenarios that could employ biological warfare agents in a terrorist or combat operation.

The spring 1995 chemical warfare attack in the subways of Tokyo is a glaring example of just how susceptible modern society is to this kind of insidious attack. It does not take a great deal of imagination to conceive of other situations and vulnerabilities that would make very lucrative targets for a biological weapons strike. If an attacker has access to the target area, a simple mechanism to aerosolize a substance, and a basic biology laboratory, the prerequisites are complete. This is not a high-tech arena that requires specialized equipment or core material as do nuclear weapons; this is basic college biology coupled with motivation. While the use of this weapon has not been prevalent in recent years, the threat is real, the United States' vulnerability is clear, and the ability to counter the biological weapon is almost nil.

A study of biological warfare raises some fundamental questions:

- Just what is biological warfare?
- What is the history of biological warfare and how important is it today?
- What biological warfare agents are available for use today?
- What is the biological warfare threat?
- How capable are we of coping with the threat?
- What policy should the United States follow to close the gap between the threat and the capability?

The purpose of this article is to raise the awareness level about a very real and probable threat that has not been dealt with effectively. The author hopes to bring the issue to the front burner for study and to apply resources to resolving the tough problems. While the paper identifies where precious resources should be focused, it does not profess to have all the answers to the very difficult biological warfare dilemma.

First, what is biological warfare in layman's terms? From a military perspective, it is the intentional use of diseases to affect an adversary's military force, population, crops, or livestock. Certainly, a terrorist biological campaign could target those same kinds of objectives, depending on the perceived purpose of the terrorist. There are two basic categories of biological warfare agents. Microorganisms are living organic germs, such as anthrax (bacillus anthrax). Second, toxins are the byproducts of living organisms, or effectively natural poisons, such as botulism (botulinum toxin) which is a byproduct of growing the microorganism clostridium botulinum.[2] These are only two examples of biological warfare agents, although these are especially prevalent and virulent examples. There are many other natural and man-made agents that have been used throughout history.

Historical Perspective

Biological warfare is not a twentieth century development; it has been an effective combat weapon for centuries. As early as 1346 A.D., Tartars held the walled city of Kaffa under siege and catapulted plague-infested bodies into the city.[3] Were the Tartars successful in using disease as a means to break the siege? Yes. Not only did illness cause Kaffa to capitulate, but some medical historians speculate this event resulted in the bubonic plague epidemic that spread across medieval Europe between 1347 and 1351, killing 25 million people.[4]

Three hundred years later, during the French and Indian War, the English offered blankets to Indians holding Fort Carillon. The English suspected the Indians were loyal to the French and exposed the blankets to the smallpox virus before their apparent altruistic overture. The Indians began to fall ill, and after an epidemic spread through the fort, the English attacked, defeating the incapacitated force. The British gained control of Fort Carillon and renamed it Fort Ticonderoga.[5]

Throughout history, many examples may be found illustrating the use of natural diseases in war to place an adversary in a position of disadvantage. For example, dumping bodies into water supplies has been fairly common for centuries. Two thousand years ago, Romans fouled many of their enemies' water sources by throwing the corpses of dead animals into the wells.[6] During the American Civil War, Confederate soldiers shot horses and other farm animals in ponds in an effort to contaminate the water supply of the Union forces.[7]

While there was some evidence of biological warfare in World War I[8], the interwar years saw a new interest in the use of disease as a weapon. Paradoxically, probably the two most active programs started as a result of an international initiative to ban biological warfare agents. Both Japan and the United Kingdom recognized that since biological warfare was horrifying enough to outlaw, it probably would make an effective weapon. Both countries had very robust programs as early as 1932 and 1934, respectively.[9]

There is evidence that Japan tested biological warfare agents on prisoners of war and that they actually used them on the population of China.[10] To spread the plague, they dropped flea-infested debris over 11 cities in mainland China. The result was a bubonic plague epidemic in China and Manchuria.[11] While these attacks caused casualties, the weapons did not function reliably and ultimately resulted in very little strategic impact that affected the war.[12]

When the Britain learned of the Japanese biological warfare program, they put significantly more emphasis toward developing their own BW capability. Most of their testing was conducted on an island called Gruinard off the northwest coast of Scotland. They concentrated their development and testing efforts on the lethal effects of anthrax. Scientists used sheep as victims to evaluate the effectiveness of the disease, and they infected literally thousands of animals. As a result of the huge amount of anthrax agent dispersed on the island and the large number of sheep infected, the British could not effectively decontaminate the island after they stopped the testing program. Consequently, Gruinard is still considered contaminated and is off limits, demonstrating the persistence of anthrax as a biological weapon.[13]

The British soon combined their biological weapons development efforts with Canada and the United States. Even though there were Allied operational plans to employ biological weapons during World War II, there is no evidence to indicate they were actually used on a large scale. There is, however, strong evidence that Reinhard Heydrich, chief of the Nazi security service, was assassinated with a grenade that had been contaminated with biological warfare agents (typhoid fever).[14]

Biological Warfare in the Cold War

After World War II and during the Korean War, the focus, at least from the United States perspective, was on building a BW retaliatory capability. The US developed an anticrop bomb and delivered it to the Air Force in 1951. It could have been used to attack North Korean rice fields, reducing a significant source of nutrition for the population.[15] North Korea accused the United States of using biological agents during the Korean War; the United States denied the accusation, and there was no substantive proof offered in the open literature.[16]

Following the Korean War, the United States invigorated the biological warfare program in 1956 after Marshal Zhukov announced to the Soviet Congress that chemical and biological warfare weapons would be used as weapons of mass destruction in future wars. This was a dramatic shift in Soviet policy and the cold war philosophy.[17] The fundamental concept of United States biological warfare operations changed as a result.

During the Korean War, the biological capability was maintained primarily for retaliation in the event an adversary employed a biological agent against United States or allied forces. The prevailing philosophy was that the threat of retaliation in kind would deter the use of these kinds of weapons. After the new Soviet pronouncement, the United States concept changed to employment upon executive order by the president of the United States.[18] Effectively, this mimicked the Soviet position, implying that the United States might use biological weapons in situations other than straightforward retaliation. This change in policy boosted the biological warfare research effort in the United States.

The bulk of the research was conducted at Fort Detrick in Maryland. It was during this "boost phase" that United States vulnerability was clearly demonstrated with simulated covert biological warfare attacks on at least three cities subway systems. Surrogate biological agents were introduced into the air vents of the underground systems. Samples were then taken to determine how widespread the dissemination would be. The results demonstrated that large numbers of the populace would be exposed to infectious doses under such an attack.[19] This experiment supported a similar test that took place in 1950 when an aerosol cloud of a simulated biological agent was sprayed off the coast of San Francisco. The test results showed that nearly 100 percent of the population had inhaled potentially lethal doses.[20]

In 1969, President Nixon changed the United States policy on biological warfare. During a visit to Fort Detrick, he announced that the United States was terminating research on biological warfare and was unilaterally disarming any capability to conduct offensive biological warfare. By 1972, the United States biological weapons stockpile was completely destroyed.[21] This gesture by the United States was the catalyst for the world community to embrace the Biological and Toxin Weapons Convention (BWC). A total of 118 countries (including the USSR and Iraq) signed up to abide by the BWC, which directs that the signatories will "never in any circumstances develop, produce, stockpile, or otherwise acquire or retain any biological weapons."[22]

During this time, the Second Indochina War was raging. However, there is no clear evidence that biological warfare agents were used in this conflict. Agent Orange, a herbicide, was a chemical-based agent that saw wide use, but biological weapons per se were not used.[23] While the United States biological warfare program was flourishing and even after our unilateral biological warfare disarmament, there is evidence that the Soviet program was thriving, although they had signed the BWC in 1972. In the late 1970s and early 1980s, there were reports that the Soviets used biological weapons in Laos, Kampuchea, and Afghanistan. While widely reported as a program called "Yellow Rain," these allegations were never proven.[24]

In 1978, Georgi Markov, a popular writer and Bulgarian exile, was walking to the BBC in London where he broadcast to his homeland from Radio Free Europe. As he was walking, he suddenly felt a sharp pain in his leg. Turning around, he confronted a man picking up an umbrella. The man apologized and went on his way. Markov took ill that night and died several days later. The autopsy found a small metal pellet coated with ricin, a biological toxic substance derived from the castor oil plant.[25]

Another incident occurred in April 1979 when a loud explosion was reported from a research compound in Sverdlovsk, USSR. Over the next few days, reports of an outbreak of anthrax surfaced. The United States claimed that the outbreak was the result of an accident in a biological warfare production plant. The USSR vehemently denied the accusations, claiming it was caused by tainted black market meat and poor hygiene among the population. In the media and technical literature before 1992, many Western journalists and scientists argued that the facts supported the Soviet claims.[26]

However, in 1992, Russian President Boris Yeltsin admitted that the Sverdlovsk incident was actually a biological warfare accident involving anthrax.[27] Thereafter, President Yeltsin signed a decree that recommitted Russia to the Biological and Toxin Weapons Convention. But in 1994, three defectors revealed an ongoing Russian biological warfare program that concentrates on a "superplague" for which, reportedly, the West has no antidote. President Yeltsin claimed he didn't know about any biological warfare programs. The defectors verified his claim and inferred that the military is running the program without Yeltsin's knowledge or consent.[28]

Biological Terrorism

In 1984, the French authorities made a startling discovery that demonstrates how vulnerable the world is to biological terrorism. The Paris Police raided a residence suspected of being a safe house for the German Red Army Faction. As they conducted their search, they found documents that revealed a strong working knowledge of lethal biological agents. As the police continued the search to the bathroom, they came across a bathtub containing many flasks filled with what turned out to be Clostridium Botulinum, the microorganism that produces botulism, one of the most lethal biological substances known to man.[29]

On 20 March 1995, the Tokyo subway system was attacked with chemical warfare agents by, allegedly, a cult called the Aum Shinri Kyo, or the Supreme Truth. This incident killed at least 11 people and injured as least 5,500 others.[30] Five different subway cars were struck simultaneously by individuals leaving canisters dispersing a Nazi-developed nerve agent called Sarin.[31] This is an exceptionally significant event because it strikes at the core of society with furtive lethal gases, exposing glaring vulnerabilities and fomenting terror among the population. As one victim of the subway attack said, "We're just innocent, ordinary people. It frightens me to think how vulnerable we are."[32]

On the 28th of March, Tokyo police also found large quantities of the biological warfare agent Clostridium Botulinum during one of several raids on Aum Shinri Kyo facilities.[33] This discovery clearly demonstrates that a terrorist organization had the resolve, the biological agent, and the wherewithal to conduct a horrendous biological attack against an unprotected population. As *Time* magazine said, ". . . garden-variety madness had got access to weapons of terror."[34]

BW and the 1991 Gulf War

These recent world biological warfare events have been alarming, but what really brought the biological warfare issue into the spotlight of the public's eye was the experience in Desert Storm, the Persian Gulf War. By the time of the Iraqi invasion into Kuwait, it was widely acknowledged that Iraq had a biological warfare program, concentrated on very toxic botulinum toxin and very resilient anthrax.[35] This assessment was derived from a compilation of several sources and indicators, the most dramatic being an Iraqi defector who was a microbiologist. He told a British newspaper correspondent that as early as 1983 Iraqi scientists were developing and testing biological warfare agents:

> There were many strains, botulism, salmonella, and anthrax. Friends told me they had found a way to make anthrax even more toxic. I know they experimented on sheep with Clostridium Botulinum type C (the source of botulinum toxin).[36]

The defector said he personally had done research and solved difficult technical problems relating to the weaponization and deployment of biological warfare agents.[37] This has since been confirmed officially by a representative of Saddam Hussein's present government. On 2 August 1990, when Iraqi army invaded Kuwait, the Iraqis had spent close to $100 million on their offensive biological warfare program and had a significant stockpile of biological warfare agents.[38]

Saddam Hussein announced "loud and clear" that this war would be the "mother of all wars," implying a no-holds-barred engagement.[39] This was the first time since World War II that the United States had faced a military adversary with a highly probable biological warfare capability and the resolve to use it.[40]

The United States was challenged not only with how to protect the military forces but how to preempt the use of Saddam's biological warfare arsenal. Plans for force protection included protective equipment and vaccinations against probable biological warfare threats.[41] In addition, planners were challenged to determine a mechanism to destroy the biological warfare stockpiles before Saddam could deploy them. Dropping a precision-guided bomb on the suspected storage bunkers would have been easy enough. The real challenge was destroying the viability or utility of the biological weapons without spreading the agents and causing massive collateral damage in terms of human lives. The military was simply not prepared for this eventuality.

Several tests were conducted over a very short time to try to find the right kind of enhanced munitions or bomb that would render the biological warfare agent unusable to the Iraqis and not release lethal agents into the atmosphere. The crash program was not fruitful. However, in the effort, computer modeling showed that the design of the suspected biological weapons storage bunkers offered a bombing approach that might inhibit the release of the agents. In the eleventh hour, this concept detailing specific fusing, type of bomb, and angle of attack was telephoned to the Central Command CENTCOM planners in Riyadh.[42] All suspected bunkers were attacked, and there was no confirmed collateral damage as a result of released biological agent. There was, however, one unconfirmed news report of several incidents of illness and death in Iraqi guards after the coalition bombed a biological warfare facility in Baghdad.[43]

In the end, it appears that Saddam Hussein did not use biological weapons during Desert Storm. While the Iraqi rationale may never be known for certain, it is likely that they were deterred by public signals like the one Secretary of Defense Dick Cheney announced during a news conference on 23 December 1990.[44]. . . Cheney said that "were Saddam Hussein foolish enough to use weapons of mass destruction, the US response would be absolutely overwhelming and it would be devastating."[45]

In an even more direct and unambiguous message, a 5 January 1991 letter President George Bush said to Saddam Hussein: "The United States will not tolerate the use of chemical or biological weapons... The American people would demand the strongest possible response. You and your country will pay a terrible price if you order unconscionable acts of this sort."[46]

After the cease fire, Iraq officialsadmitted having a biological warfare program that they said had only progressed to the research stage. Inspectors found evidence of a robust biological warfare production capability, but could not specifically link it to the biological warfare program.

However, there was overwhelming circumstantial evidence that an offensive biological warfare production and weaponization program did exist.[47] Like the Soviet Union, Iraq had previously signed the BWC. The bad news is that United Nations inspectors were not able to locate Saddam's biological stockpile.[48] Saddam's representatives have since admitted to the United Nations inspectors that they had a sizable cache of anthrax and botulism agents, but they claim to have destroyed it to avoid having germs spread over the Iraqi countryside by allied bombing attacks.

Having witnessed the bold Iraqi deceptive effort regarding their nuclear research program, the world has every reason to believe that Saddam Hussein still has a large amount of biological warfare agents at his disposal today.[49] On 24 February 1993, former CIA Director James Woolsey told the Senate Governmental Affairs Committee: "Iraq's biological weapons capability is perhaps of greatest immediate concern. Baghdad had an advanced program before Desert Storm, and neither war nor inspections have seriously degraded this capability. The dual-use nature of biological weapon equipment and techniques makes this the easiest program to hide."[50]

The Biological Warfare Threat

With the public exposé of active Russian and Iraqi biological warfare programs, the threat of these weapons looms large on the horizon. There are official, open-source estimates that between 10 and 20 countries either have, want, or are thinking about starting a biological weapons capability.[51] However, there is more to the threat than just countries that have the capability. What types of agents are a threat and how will they mature given new technology? And, does the insidious nature of biological agents pose a threat?

BW Nation States

Some of the countries suspected in open sources of having or wanting a biological warfare program include Russia, Syria, Iraq, Iran, Libya, North Korea, Israel, Egypt, Cuba, Taiwan, China, Romania, Bulgaria, Pakistan, India, and South Africa.[52] There are real concerns with this list. First, some of these nations have been associated in the past with state-supported terrorism. This fact raises the probability of a biological warfare terrorist attack.

Second, many of these countries reside in regions of historical instability or emerging instability. And third, with the economic distress in the former Soviet Union, there is a possibility that its biological warfare weapons experts will look for more prosperous employment by building biological warfare programs elsewhere for the highest bidder. Fortunately, as of early 1994, the CIA had no indication that this biological warfare brain drain is occurring.[53]

Biological Warfare Technology

The degree of sophistication of each country's research program will determine how advanced biological agents will be.

Even the most rudimentary program will likely have lethal agents that have been a threat for some time. Botulism and anthrax (mentioned earlier) are high-probability candidates that are difficult to reckon with. In addition, the revolution in biotechnology may produce other agents that are even more toxic and resilient. Without getting into the technical aspects, relatively minor molecular adjustments may produce a more toxic, fast acting, and stable biological agent.[54]

There is also a possibility that genetic engineering may produce a weapon that is unique and can only be protected against with a unique vaccine.[55] These two examples of potential developments in biological warfare will give this weapon a great deal more utility, especially on the battlefield. A more stable agent that produces an accelerated reaction would provide the tactical commander with a viable tactical weapon.

Additionally, if the commander could deploy biological agents against an enemy while friendly troops remained invulnerable, the biological option would become much more attractive as a battlefield weapon. There is also some speculation that a toxic agent could be produced that would target only a specific genetic makeup, giving an attacker the capability to discriminate among age, gender, racial or behavior groups as target sets.[56] Following the Tokyo subway attack, it has come to light that the Aum Shinri Kyo had recently ordered sophisticated molecular design software. The purpose of this type of software is to reengineer the molecular structure of chemicals or microorganisms to make them stronger or more dangerous.[57] Could it be that this fanatic cult was planning to use this software to genetically reengineer their biological or chemical agents?

Stealthy BW

Now the really sobering part—biological warfare agents are very difficult, if not impossible, to detect while they are in the research, production, transit, or employment phases. Normal biological warfare research facilities resemble completely legitimate biotechnical and medical research facilities. The same production facilities that can produce biological warfare agents may also produce wine and beer, dried milk, food, and agricultural products.[58] The challenge this presents is in distinguishing legitimate production plants from illicit ones.

It becomes nearly impossible to identify the locations and facilities that are actually producing biological warfare weapons. This needs to be done, obviously, in order to confidently highlight a violation of the BWC, or, if necessary, should all peaceful remedies fail, preemptively strike a biological weapons production or storage facility.

In addition, biological warfare agents are virtually undetectable while they are in transit. In other words, if a terrorist wanted to carry the biological agent into the United States in a carry-on bag or checked luggage, there is no mechanism using routine customs, immigration, drug scan, or bomb search procedures to identify the agent. The only way to find it would be a physical search by a very well trained and very lucky searcher.[59] Similarly, the threat on the battlefield is almost as insidious, with very little present detection capability.

Desert Storm represents a recent experience in which the United States needed the ability to detect biological warfare agents to give early warning for protective measures. With few exceptions, the capability was not there. The limited capability that was deployed was the result of a crash program to produce a biological detector—it was an experiment.[60] It seems logical that the inability to detect and thereby protect the civilian population or military force would significantly add to the viability of biological weapons as a terrorist or tactical battlefield threat.

Shortfalls

In addition to the detection shortfall, the United States is unable to effectively protect the military forces (medically and nonmedically), conduct an effective preemptive counteroffensive strike, or protect the population against a terrorist attack. Given the wide spectrum of kinds of agents that make up the biological warfare threat, medical prophylactic measures (primarily vaccinations) are inadequate, and it appears they will be so at least for the near-term.[61] Personal protection in a biological warfare environment currently depends on protective clothing—the chemical warfare suit. In Desert Storm, the chemical warfare suit was adequate if fitted properly (a frequent problem) but unsuitable if worn for long durations or while in hot weather.[62]

Desert Storm also highlighted the shortfall in the ability to strike a biological warfare storage facility with confidence that massive numbers of innocent civilians would not be killed (collateral damage) as a result.[63] The United States is impotent to prevent a biological warfare terrorist attack against the population unless there is specific intelligence to forewarn of the attack.[64] Additionally, following a biological warfare attack, there are many agents that medicine can't treat today.[65]

Given this discouraging information, the scenario described in the prologue seems even more plausible. Other "concepts of operations" are not hard to imagine. Nearly every grocery or drug store sells small aerosol deodorizers that periodically spray a fragrant mist. If an adversary wanted to neutralize the military brainstem of the United States, they might refill these deodorizers with a biological agent and clandestinely place one in each restroom in the Pentagon. After a few days, the entire population of the Department of Defense headquarters would be incapacitated, causing mass confusion and widespread terror.

In a combat environment, conventional dispersal with bombs, artillery, or even a spraying device on an aircraft (like a crop duster) would not be nearly as effective as a more surreptitious attack that would infect people before they donned protective clothing. An infiltration by special operations forces or undercover operatives to place aerosol canisters similar to insect bombs or deodorizers might cripple a force before it knew it was attacked. Like the Indians at Fort Ticonderoga, the force would fall ill and many would die. The force's ability to conduct effective combat operations would certainly be negated. By the time doctors diagnosed the disease and determined the right antidote, if there were one, the war could have been lost.

Consider the implications if the Aum Shinri Kyo had used botulinum toxin or Anthrax instead of the Sarin chemical agent in their attack on the subway system in Tokyo. The death count and the magnitude of the terror would have been higher by orders of magnitude. There may have been as high as a 90 percent fatality rate instead of 0.2 percent actually experienced—that could be nearly 5,000 dead innocent civilians! And considering that the volume of Sarin to saturate a given area is approximately equivalent to 10,000 times the amount of botulinum toxin needed to cause the same effect, the attack could have been vastly more devastating.[66]

In another recent real-world incident, consider how much more effective the terrorist bombing of the New York World Trade Center would have been if they had placed a fire extinguisher filled with a biological agent at the bottom of each stairwell and rigged them to begin spraying just as the bomb ignited. In the ensuing panic, thousands of occupants of the building escaped down the stairs. No one would consider a fire extinguisher out of the ordinary in a crisis situation after the bombing. As a result, potentially every occupant in the World Trade Center could have been infected.

If the intent of the terrorists had been to demonstrate how vulnerable the population of the United States is, the addition of biological agents to the conventional attack would really have terrified leaders and other citizens in the United States.

These incidents of potential biological terrorism must raise concern and questions about civilized society's ability (or more accurately, inability) to deal with such an eventuality. As we enter the twenty-first century, we may well be facing weapons of mass destruction used, not on the battlefield by warriors but among dense population centers by deranged non-nation states—a sobering prospective. Clearly, more has to be done to overcome this dramatic vulnerability—and soon.

Resolution

Biological terrorism is a challenge for the diplomatic, technical, military, medical, and intelligence communities, but the political arena may hold the biggest stick to deter biological warfare aggression. The BWC is the international vehicle to prevent biological proliferation. Unfortunately, it does not provide for verification or punitive measures.[67] With the blatant violations of Russia and Iraq, much tougher verification protocols and stronger teeth must be built into the BWC. This is especially challenging given that the dual-use technology that produces biological agents gives the biological warfare producer an almost built-in plausible deniability.

The technical community has the greatest and most urgent challenge to develop effective detectors, both on the battlefield and in biological agent detectors similar to metal detectors. This effort should be a top priority. There should also be technological exploration, in concert with the intelligence community, for means to detect clandestine biological production facilities. The state-of-the-art must be pushed to find some means to detect a production facility with certainty, no matter the size. Both human intelligence and the national technical means must be greatly improved.

The military challenge is to train and equip to respond to a detected biological threats. To respond on the battlefield, militaries must develop effective, comfortable, and long-wearing protective clothing to replace the existing ensemble. A self-contained, air conditioned unit would be ideal. The military must also be capable of responding to a more strategic biological warfare threat—the production facilities and stored munitions. Planners must work with the technology community to develop a capability to bomb a biological warfare target and destroy the viability of the agents before they can be brought to bear on friendly forces and without causing unacceptable levels of collateral damage. For obvious political reasons, such precision-guided munitions should, also, be kept non-nuclear. The military also should hone its special operations, direct action skills for the biological (as well as chemical and nuclear) mission. The special operation option may be a more plausible alternative, depending on the scenario.

The medical community should continue to work on biological warfare vaccinations that are broad-based, safe, and in sufficient quantities to inoculate those people most susceptible to biological warfare attacks. This daunting task will be even more challenging given the controversy about the vaccines administered during Desert Storm and their suspected connection with the Gulf War Syndrome.[68] Doctors should also strive to improve the post-attack treatment in terms of rapid diagnosis, effective medical treatment, and a responsive surge capability to administer to large numbers of biological warfare-exposed patients.

The intelligence community must be strengthened and sensitized in its efforts to gather data on the biological warfare threat. More resources should be directed toward identifying biological warfare threats by human and national technical means. This is especially important to deter terrorism in the interim until human intelligence and national technical means can provide more definitive answers about who are the haves and have nots.

Finally, United States and allied political leadership should articulate a clear retaliatory policy against the use of any weapon of mass destruction. This was an effective deterrent on both sides during the cold war, and it appears to have deterred Saddam Hussein during Desert Storm. Perhaps even more importantly, this policy must be supported by unrelenting resolve to actually carry out the retaliation.

Since countering BW is an issue that crosses many government agency borders, the direction of the effort should come from a multi-agency steering group. This steering group initially should include principals or primary deputies from the Office of the White House, Department of Defense, Federal Emergency Management Agency, Public Health Service, Central Intelligence Agency, and Department of Justice. Well-versed technical and operational advisors will be essential to steer the effort.

Many of these agencies already have ongoing programs, but there is little senior-level cohesion to these fragmented endeavors. Additionally, some of these efforts have demonstrated blatant parochialism. A multi-agency steering group would overcome these stovepipe attitudes and efforts, placing emphasis on national interests and prioritizing accordingly.

Conclusions

Biological warfare has been a threat for decades if not centuries. Yet the United States is ill-prepared to defend against or counter it—why? One view is that "the United States has a tendency to wish the problem would go away because it seems too unsavory and too difficult to handle."[69] Another skeptic says, "We don't need it [BW defense] because we have a treaty."[70] It seems the real issue is the apparent imbalance between demonstrated threat versus resources expended to meet that biological warfare threat.

In the case of biological warfare, the fixes are technically difficult and they will not be low-cost. Weigh this against a threat that has not yet fully manifested itself. It almost seems logical that decision makers would be reluctant to spend scarce resources against a heretofore invisible threat. However, the United States is moving toward a more aggressive counter-BW program. In February 1995, the White House published a national security strategy that said:

> U.S. forces must be prepared to deter, prevent and defend against their use. The United States will retain the capacity to retaliate against those who might contemplate the use of weapons of mass destruction, so that the costs of such use will be seen as outweighing the gains. However, to minimize the impact of proliferation of weapons of mass destruction on our interests, we will need the capability not only to deter their use against either ourselves or our allies and friends, but also, where necessary and feasible, to prevent it. We are placing a high priority on improving our ability to locate, identify, and disable arsenals of weapons of mass destruction, production and storage facilities for such weapons, and their delivery systems. To minimize the vulnerability of our forces abroad to weapons of mass destruction, we are placing a high priority on improving our ability to locate, identify and disable arsenals of weapons of mass destruction, production and storage facilities for such weapons, and their delivery systems.[71]

This is a step in the right direction, but it needs to be a giant step. The biological warfare threat looms. The United States must have the capability to detect, preempt, and protect before someone strikes us or our allies with a poor man's nuke.

Epilogue—9 January

This is a CNN special report live from the Anthrax Task Force Center Miami. This morning, the fatality count was 16,437. This grim figure was just given to us by doctors here. Unfortunately, they say the number is going to increase dramatically because so many patients are close to death right now. Doctors are working frantically to save as many as possible, but they are running out of antibiotics and facing massive overcrowding. The halls are crowded with gurneys, and relatives are being asked to wait outside unless their loved one is critical. And there are many of those.

The Anthrax Task Force was quickly assembled on the sixth of January after doctors across the nation diagnosed the horrible epidemic as pulmonary anthrax. The Federal Emergency Management Agency heads the team that consists of representatives from the FBI, the Center for Disease Control, the Armed Forces Military Intelligence Center, and the US Army Research Institute of Infectious Diseases, to name just a few. They are warning anyone who attended the Orange Bowl on New Year's Day to seek medical attention immediately. If you are experiencing cold-like symptoms, you are probably infected. Do not hesitate, or it will be fatal. The FBI reports that this appears to be a deliberate act of mass murder. But that is all they have been able to determine. They are offering a ten million dollar reward for any information about this horrendous crime.

This is all from Miami. Back to CNN News Headquarters.

Notes

1 *Montgomery (Alabama) Advertiser*, November 24, 1994, sec. B.

2. Kathleen C. Bailey, ed., *Director's Series on Proliferation* (Springfield, Va.: Lawrence Livermore National Laboratory, May 23, 1994), 2.

3. Robert Harris and Jeremy Paxman, *A Higher Form of Killing* (New York: Hill and Wang, 1982), 74.

4. Ibid., 9.

5. Bailey, 9–10, and Harris and Paxman, 74.

6. Charles Piller and Keith R. Yamamoto, *Gene Wars, Military Control Over the New Genetic Technologies* (New York: Beech Tree Books, 1988), 29.

7. Ernest T. Takafuji, M.D., M.P.H., Col, US Army, "Biological Weapons and Modern Warfare" (Fort McNair, Washington, D.C., The Industrial College of the Armed Forces, National Defense University 1991), 4.

8. Bailey, 10.

9. Harris and Paxman, 75–81.

10. Sheldon H. Harris, *Factories of Death, Japanese Biological Warfare 1932–45 and the American Cover-up* (New York: Routledge, 1994), 113–131.

11. Bailey, 10–11, and Harris and Paxman, 75–83.

12. Jonathan B. Tucker, "The Future of Biological Warfare," in Thomas Wander and Eric H. Arnett, eds., *The Proliferation of Advanced Weaponry* (Washington, D.C.: AAAS, 1993), 16.
13. Harris, *Factories of Death*, 68–74.
14. Ibid., 88–93.
15. Bailey, 15.
16. Harris, *Factories of Death*, 162–63.
17. United States Army, *U.S. Army Activity in the U.S. Biological Warfare Programs*, vol. 2 (Washington, D.C.: US Government Printing Office, 1977), 29.
18. Bailey, 16, and Harris, *Factories of Death*, 164.
19. Ibid, 17.
20. Tucker, 18.
21. Bailey, 19, and Jeanne McDermott, *The Killing Winds, The Menace of Biological Warfare* (New York: Arbor House, 1987), 30–32.
22. Harris, *Factories of Death*, 171–72, and Tucker, 6–7.
23. McDermott, 189–90.
24. Leonard A. Cole, *Clouds of Secrecy, The Army's Germ Warfare Tests over Populated Areas* (Landam, Md.: Rowman & Littlefield, 1988), 107–19, and Stephen Rose, "The Coming Explosion of Silent Weapons," *Naval War College Review*, 42, (Summer 1989): 26–27.
25. Bailey, 11; Harris, *Factories of Death*, 197–98; and McDermott, 156.
26. Cole, 131; Harris, *Factories of Death*, 220–22; and McDermott, 37–44.
27. Bailey, 12
28. "Report: Russia making 'Superplague Germs," *The Honolulu Advertiser*, 28 March 1994.
29. Joseph D. Douglas, *America the Vulnerable: The Threat of Chemical/Biological Warfare, The New Shape of Terrorism and Conflict* (Lexington, Mass.: Lexington Books, 1987), 29.
30. *Montgomery Advertiser*, 8 April 1995.
31. David Van Biema, "Prophet of Poison," *Time*, 3 April 1995, 28.
32. Ibid., 29.
33. Mari Yamaguchi, "Japanese Cult Had Bacteria Useful for Germ Warfare" *San Francisco Chronicle*, 29 March 1995.
34. Van Biema, 28.
35. United States Department of Defense (DOD), Office of the Secretary of Defense, *Conduct of the Persian Gulf War, Final Report to Congress* (Washington, D.C.: US Government Printing Office, April 1992), 15.
36. Jonathan B. Tucker, "Lessons of Iraq's Biological Warfare Programme" *Arms Control*, December 1993, 237.
37. Ibid.
38. Ibid., 240.
39. DOD, 639.
40. Tucker, 2.
41. Bailey, 27.
42. This is based on the author's experience as chief of the team chartered with finding the solution to attacking biological warfare storage and production facilities.

43. "Biological Plan Bombed," London: *Reuters Wire Service*, 3 February 1991.
44. John M. Broder and David Lamb. "Some Allies Won't Join Offensive, Cheney Says." *Los Angeles Times*, 24 December 1990.
45. DOD, 639–40.
46. President George Bush, "Crisis in the Gulf," *US State Department Dispatch*, II, 2, (14 January 1991): 25.
47. Tucker, 250–59.
48. Bailey, 26–27; and Rose, 11–12.
49. Defense Nuclear Agency, *Biological Weapons Proliferation* (Ft. Detrick, Md.: US Army Medical Research Institute of Infectious Diseases, April 1994), 49.
50. Tucker, 259.
51. Defense Nuclear Agency, 46.
52. Ibid.
53. Ibid., 50.
54. Rose, 29.
55. Tucker, 12.
56. John F. Guilmartin Jr., and Sir Michael Howard, *Two Historians in Technology and War* (Carlisle Barracks, Pa.: US Army War College, 20 July 1994), 33.
57. Prodigy Services Company. Software Returned to U.S. Firms, 3 April 1995.
58. Defense Nuclear Agency, 50.
59. Rose, 35–46.
60. Ibid., 82.
61. Bailey, 67–76.
62. DOD, 639–46.
63. Rose, 85.
64. Ibid., 35–46.
65. Bailey, 62–63.
66. Ibid., 40–41.
67. Ibid., 85–90.
68. *USA Today*, December 8, 1994.
69. Rose, 20.
70. Bailey, 35.
71. The White House, *A National Security Strategy of Engagement and Enlargement* (Washington, D.C.: Government Printing Office, February 1995), 14.

Chapter 9

Twenty-First Century Germ Warfare

Lt Col Robert P. Kadlec, USAF

The United States military is entering "one of those rare historical periods when revolutions happen in how wars are fought. The revolution derives not from any single invention or idea, but from a range of rapidly developing technologies."[1] Some ten military revolutions have occurred since the fourteenth century.[2]

Advances in sensors, communications, stealth technology, and precision munitions have preoccupied those leaders planning how the United States will wage future wars. The revolution in biotechnology, however, has gone relatively unnoticed. The same technology and expertise which has brought revolutionary medical therapies and greater agricultural productivity is readily transferable to the development of biological weapons.

Many technical barriers that once limited the effective use of biological warfare (BW) are gone. A country or group with modest pharmaceutical expertise can develop BW for terrorist or military use. As the United States prepares itself for the national security challenges of the twenty-first century, it must grasp the implications of this silent revolution.

Nature has long waged its own form of biological warfare. The epidemic of bubonic plague killed an estimated one quarter of Europe's medieval population (25 million deaths) between 1347 and 1351. The introduction of smallpox into the New World by European explorers decimated the population of Native Americans.[3] A pandemic of Spanish flu may have killed 50 million people worldwide between 1918 and 1919.[4] By the year 2000, 40 million people could be infected by the Human Immunodeficiency Virus (HIV) which causes the Acquired Immunodeficiency Disease Syndrome (AIDS).[5]

Such events as these undeniably change biologic, economic, and political systems. Governments, groups, and individuals who desire weapons of mass destruction (WMD) can use biotechnology to achieve this goal. Skeptics mistakenly dismiss the military or strategic value of biological weapons. These weapons represent a credible threat to United States security and future economic prosperity.

Biological warfare offers an adversary unique and significant advantages because of its ease of production, potential impact of use, and the ability to exploit US vulnerabilities. It is the only weapon of mass destruction which has utility across the spectrum of conflict. Using biological weapons under the cover of an endemic or natural disease occurrence provides an attacker the potential for plausible denial. In this context, biological weapons offers greater possibilities for use than do nuclear weapons.

Biological warfare can include the use of bacteria, rickettsia, viruses, and toxins to induce illness or death in humans, animals, and plants. In the current public opinion, there is a significant misperception that clouds BW discussions. Biological warfare is often lumped together with chemical weapons. In BW, the types of agents, physiologic effects, methods of protection and detection, and methods of application are distinctly different from those of chemical warfare (CW).

Chemical Warfare Versus Biological Warfare

Biological agents are many times deadlier, pound-for pound, than chemical agents. Ten grams of anthrax spores could kill as many people as a ton of the nerve agent Sarin.[6] There are four distinct types of chemical weapons: nerve, blister, blood, and incapacitating agents. The effects from these chemical agents can occur within seconds of exposure as in the case of nerve and blood agents or as long several hours in the circumstance of low-dose blister agent exposure such as mustard gas. The physiological and medical effects of CW are limited to well-defined symptom complexes. The outcome of each exposure is dose-dependent death or incapacitation.

Of the four general types of biological warfare agents mentioned, 60 have been identified with potential weapon utility against humans.[7] The medical effects of biological agents are diverse and are not necessarily related to the type of agent. Some cause pneumonia. Others can cause encephalitis or inflammation of the brain. Each one causes a different complex of symptoms, which can either incapacitate or kill its victim.

The ineffective dose required to induce illness or death may be as great as tens of thousand of organisms as in the case of anthrax, or just a few as with tularemia. With the exception of exposure to a toxin, a period of several days or even weeks may pass before the onset of symptoms and the ultimate effect. This incubation period is the time necessary for the microbe or viral agent to establish itself in the host and replicate.

Toxins, on the otherhand, are a product of living organisms and behave similar to chemical agents. Botulinum toxin is the most toxic substance known to man. Without supportive care, inhalation of nanograms (10^{-9} milligrams) of this agent will cause progressive muscular paralysis leading to asphyxiation and death.

Hazards and Protection

The risk posed by chemical agents has two components: a vapor and liquid hazard. An individual is protected from the vapor and liquid hazard of CW with the combination of a protective mask and the Mission Oriented Protective Posture (MOPP) suit. The United States also possesses an array of chemical point detectors and alarms which provide real time warning of exposure or attack.

In contrast, the hazard posed by biological agents is primarily an inhalational one. The most effective means of delivering a BW agent is via an aerosol in the one to five micron particle size. Creation of this type of an invisible aerosol cloud may be efficiently accomplished using an agricultural sprayer for example.

The current US military protective mask, when properly fitted and donned, affords virtually 100 percent protection. Biological agents generally do not pose a cutaneous hazard. A MOPP suit is not required. When US troops deployed to Saudi

Arabia during Operation Desert Shield, they did not have an operational capability to detect any biological agents.[8]

During Desert Storm, the United States fielded only a rudimentary developmental detector system. This system could detect only two of several possible Iraqi BW threats. In addition to the limited scope of this detector, it took between 13 and 24 hours after the attack to determine the presence and identify the BW agent. There was no capability to provide any real-time or advanced warning of a biological attack.[9] During the Gulf War, the first likely indication of an attack was ill and dying soldiers.

Biological Warfare Detection and Medical Protection

Detection of biological agents is a complex problem. Since World War II, attempts by the United States to develop BW detection have met with frustration and only limited success. Given the chemically indistinguishable organic properties of biological agents, the methodology for detecting chemical agents is not useful by itself. Each prospective BW agent requires a specific assay to detect and identify. Advances in medical diagnostics and biotechnology are allowing science to overcome these technical obstacles. The number of potential agents, however, and the demanding technical and developmental requirements make the challenge of BW detection daunting.

In lieu of detection or advanced warning and the opportunity to don protective masks, medical products such as vaccines, immunoglobulins and antibiotics can mitigate the effect of biological agents and the potential operational impact. All three types of products can provide both preexposure and postexposure protection to an infectious disease and therefore a BW agent.

Vaccines are active immunization measures whereby the host is intentionally exposed to either an attenuated or inactivated form of the disease causing agent. Vaccines prompt the body to produce antibodies which can afford a high level of protection for many years. The current Food and Drug Administration (FDA) approved vaccine against anthrax has been shown to be protective against inhalational agents in nonhuman primates.[10]

Like biological weapon detectors, vaccines are highly specific for a potential threat agent. In addition, the time needed to develop a safe vaccine product suitable for human use may be 10 to 15 years. While several BW vaccines exist for known threat agents such as botulinum toxin and tularemia, they remain in a status short of full FDA approval known as an investigational new drug (IND).

Full FDA approval may be slow in coming if there is not sufficient test data on human response to the vaccines or if there is a lack of commercial demand or no perceived near-term public requirement for its use. Vaccines against exotic infectious disease agents associated with offensive BW fall under the purview of the Department of Defense with the Army as research and development executive agency.

The ability to provide IND products, which are shown to be safe in humans and efficacious in animal studies, short of war or national crisis is limited by the need to satisfy the human use or informed consent requirements. These requirements specify the need to inform the prospective recipient of possible known and theoretical, short-term and long-term effects of the product. The stigma attached with IND status may have limited the use of the vaccine during Desert Storm. The controversy surrounding Persian Gulf Illnes (PGI) further negatively effects perceptions of IND products used for military purposes. To date the biological warfare vaccines have not been associated with PI.

Once a vaccine is given, there is lag time before adequate protective antibodies develop. The length of time and the number of doses of vaccine required before protection is attained and its duration, are unique characteristics of the specific product. Furthermore, a sufficiently high dose of a BW or infectious disease agent can overwhelm any vaccine. Finally, the protective level obtained for a population is not uniform. There is individual variability.

Immunoglobulins are existing antibodies which can be harvested from humans or animals. Like vaccines, they are protective for a specific agent. Unlike vaccines, which stimulate active production of antibodies by the host, immunoglobulins are considered passive immunization. When injected into a host, they provide short-term protection which usually lasts weeks or months. Immunoglobulins are usually used to protect viral and toxin agents. The rigorous FDA approval required for vaccines is also required for immunoglobulins.

A well-known example of an immunoglobulin preparation is the serum used for Hepatitis A. Immunoglobulins currently exist for pre-exposure and post-exposure treatment of botulinum toxin intoxication. Developments in monoclonal antibody technologies offer the potential for rapid development of immunoglobulin preparations for use in BW medical defense.[11]

Finally, the use of antibiotics for pre-exposure or post-exposure treatment of bacterial or rickettsial BW agents offers the potential for broad spectrum-agent protection. Antibiotics however do not prevent or treat illness due to viral or toxin agents. There are further practical theoretic limitations to the use of antibiotic. Certain microbial infectious diseases and BW agents have a natural resistance to different antibiotics. Furthermore, antibiotic resistance among microbes occurs naturally or can be developed deliberately. Deliberate resistance can be induced by exposing cultures to serial exposures of ever-increasing doses of an antibiotic or by transferring a piece of genetic material which confers antibiotic resistance.[12] It is possible to purposefully develop a BW agent that is resistant to a range of different antibiotics.

The duration of protection afforded by antibiotics depends on the amount of antibiotic available and the compliance of the individual to take them. For certain bacterial BW exposures, antibiotics may have to be taken for more than 30 days to prevent illness and possible death.[13] A large enough infective dose of a bacterial or rickettsial agent could, as in the case of the vaccine, overwhelm the protection afforded by antibiotics.

The medical means to mitigate the effects of BW agents and attack are imperfect. Even with reliable detectors which provide real time or advanced warning, vaccines, immunoglobulins, and antibiotics play a complementary role in the passive defense against BW. In the absence of BW detection, medical products represent the cornerstone of protecting US forces from the effects of BW. Besides their medical benefit, their administration may deter an enemy from using BW.

Role of Intelligence and Medical Defense

The need for a prior knowledge of BW agent threats for both detectors and medical products has already been suggested. A great deal is known about US offensive BW agents developed and produced until 1969 (Table 2).

Table 2

US OFFENSIVE BIOLOGICAL WEAPONS AGENTS PRODUCED BETWEEN 1954–1969

Antipersonnel	Anticrop
Anthrax	Wheat rust
Tularemia	Rye rust
Brucellosis	Rice blast
Q Fever	
Venezuelan Equine Encephalitis (VEE)	
Botulinum Toxin	
Staphyloccal	
Enterotoxin	

There are some classical BW agents which Japan, the United States, and possibly the former Soviet Union and Iraq have researched and developed as weapons.[14] Anthrax and botulinum toxin fall into this category. Among the 60 other potential BW agents, "sound and complete intelligence data" will play a vital role in ascertaining which ones proliferants will pursue.[15] Robert Gates, then director of the Central Intelligence Agency (CIA), underscored the problem in January 1992, when he cited human intelligence as being critical in assessing the proliferation of both chemical and biological weapons in the third world.[16]

Congress has not missed the importance of intelligence in focusing R&D. Congressional law mandates that BW defensive medical efforts be funded and prioritized based on a threat list developed by the US intelligence community.[17]

Two questions emerge from this process. The first is whether adequate priority and resources have been focused on collecting and analyzing BW-related intelligence and how successful these efforts have been. The second is whether the process of creating a threat list based entirely on intelligence assessments is adequate. Assessing an adversary's current BW efforts may ignore developments in biotechnology which may offer breakthrough capabilities and novel agents.

Weaponizing BW

The final distinction between chemical and biological weapons involves the methods of dissemination. Weaponizing of CW and BW implies three essential elements: agent, munition, and delivery system. Both can be "weaponized" into conventional munitions such as artillery rounds, cluster bombs, and missile warheads.

Unlike CW, BW weaponizing implies the efficient aerosolization of viable agents in a one-to-five micron particle size. Dissemination of BW agents by conventional munitions pose certain technological difficulties, because they are sensitive to environmental stresses. Excessive heat, ultraviolet light, humidity and oxidation decrease their potency and persistence.[18] The United States overcame these difficulties in the late 1960s to produce, weaponize, and stockpile several BW agents.[19]

Weaponizing BW includes some unconventional delivery systems which may put the United States at risk.[20] In contrast to chemical agents, biological agents could also be easily adapted for use with commercially available agricultural sprayers. This dissemination method lends itself to both covert and clandestine applications and possible terrorist use.[21] Unmanned remotely piloted vehicles (RPVs) with spray tanks represent another low-observable means of BW delivery.[22]

A common characteristic shared by both chemical and biological agents is their dissemination on ancient meteorological conditions. The prediction and control of the environmental dispersion of BW agents represents the greatest uncertainty about their use. Ideal conditions would occur at night, with favorable mild to moderate winds. The relative coverage of 1,000 kilograms of nerve agent Sarin is 7.8 square kilometers under these meteorological conditions. Attacking a major metropolitan city like Washington, D.C., would result in an estimated 3,000 to 8,000 deaths. A similar attack using 100 kilograms of anthrax under the same conditions would cover 300 square kilometers and result in 1 to 3 million deaths.[23] Anthrax, under favorable meteorological conditions, could kill as many people as a comparably sized nuclear device.[24]

The Current Biological Warfare Threat

A recently published Office of Technology Assessment (OTA) document, *Proliferation of Weapons of Mass Destruction: Assessing the Risks*,[25] had three major findings.

1. The states most actively working to develop WMD, although limited in number, are for the most part located in unstable parts of the world—the Middle East, South Asia, and the on Korean Peninsula.
2. WMD proliferation poses dangers to all nations. It poses particular problems for the United States.
3. The breakup of the Soviet Union presents immediate threats to the global nonproliferation regimes.

The proliferation of nuclear, biological, and chemical weapons and ballistic missiles creates a pall over the potential achievement of a stable global environment. It also raises the risks of escalation of regional conflicts. Proliferants understand the value of these weapons for deterrence, coercion, and war fighting.

The consequences of their use have unfortunately been recorded in recent history. The horrific images of Hiroshima and Nagasaki leave little to imagination regarding the death and destruction that occur with a nuclear detonation. Similarly, the television images of dead Kurdish villagers and incapacitated Iranian soldiers during the Iran-Iraq War reveal the grisly and inhuman effects of chemical weaponry. The psychological impact of Iran's Scud missile attacks on Israel and Saudi Arabia was enormous.

Nowhere in recent history, however, has the use of BW been similarly documented. The suspicion that "Yellow Rain" or the mycotoxin tricothecene was used in Indochina has never been conclusively proven, to the embarrassment of the Reagan administration. Similar suspicions have proven inconclusive during the Soviet Union's occupation of Afghanistan. The former Soviet Union has only recently admitted the truth about an accident at a biological agent production facility in Sverdlovosk in 1979. Boris Yeltsin, himself, recently disclosed Russia's violation of the Biological Weapons Convention in 1992.

During the 1990–1991 Gulf War, Iraq was suspected of having an extensive BW capability, but subsequent UN inspections failed to find any definitive proof. However, under the pressure of economic sanctions, Iraq admitted that it had a biological warfare R&D effort. For example, Iraqi officials have now disclosed that between 1985 and 1991 Iraq produced two germ warfare agents, baccilus anthracix and botulinum toxin.[26]

In some measure, BW has attained the stature of the "bogey man of WMD"—terribly feared but never seen. That fear and the threat, however, are real. According to the commander of the US Army Chemical and Biological Defense Agency (CBDA), the biological threat has been recently singled out as the one major threat that still inflicts catastrophic effects on a theater-deployed force.

Desert Storm solidified the perception in the United States, in the Congress, and among our military leadership that biological weapons were something that third world nations considered a potential equalizer.[27] Had Iraq used anthrax or botulinum toxin, enormous casualties would have resulted which would have overtaxed the Army's theater medical system.[28]

Certain findings of the OTA assessment relating to BW are noteworthy. The ease and cost of developing and producing BW is much simpler and cheaper than developing nuclear weapons. Biotechnology allows small facilities to be capable of producing large amounts of biological agents. Ten million dollars allows a proliferant to produce a large arsenal.[29] The scientific and technological knowledge needed to develop and produce offensive agents in significant quantities is readily available and relatively unsophisticated. The equipment required is widely available and is dual-use, having legitimate commercial applications. Finally, and probably most importantly, the use of BW could be difficult to prove in some cases since outbreaks of endemic or naturally occurring disease happen.[30]

Eight nations have been implicated in developing offensive BW capabilities: Iran, Iraq, Israel, North Korea, China, Libya, Syria, and Taiwan.[31] A ninth, Russia, admitted to developing an offensive program in violation of the Biological Weapons Convention, although it has reportedly ended such activities. The aforementioned list may not be all inclusive. Given the ability to produce militarily significant quantities of BW agents (kilograms) in small legitimate facilities, a committed proliferant with a "modestly sophisticated pharmaceutical industry" could develop a credible undetected clandestine offensive BW capability.[32]

Genetic engineering is not expected immediately to herald the development of new or exotic BW agents. Instead, its impact may enhance the environmental stability of existing BW agents. It has also been speculated that cloning DNA segments from toxin-producing organisms may allow for the mass production of these agents. Production of protein molecules such as human insulin has been demonstrated and is commercially available technology.

Nonproliferation and Counterproliferation Policy

The fall of the Soviet Union signaled the end of the superpower rivalry and created an opportunity to establish an international system that seeks to ensure political stability, economic opportunity, and collective methods of conflict resolution. The United States has prioritized its national security objectives to enhance economic growth and development; to prevent the proliferation of nuclear, chemical, and biological weapons; and to resolve regional conflicts.

Nevertheless, among aggressive third world states, WMD proliferation is occurring, and there are strong incentives to pursue BW versus nuclear or chemical weapons. Proliferants obtain significant capabilities with minimum economic costs and political risks. The 1991 Gulf War highlights this point.

The United States and United Nations claimed after the war that Iraq had an advanced BW program, although public proof was lacking and Iraq initially denied this. During the war, Iraq capitalized on the coalition bombing of a "baby milk facility" for propaganda purposes. Despite US official claims that the Abu Gharyb infant formula plant was a dual-use facility capable of producing biological weapons, CNN's Peter Arnett's visit to the site and report raised public doubts about whether this was a legitimate allied target, when, in fact, it was. Following the war, intrusive on-site visits by UN inspectors failed to uncover any irrefutable evidence of an offensive BW program, although the Iraqis later admitted they had procured large quantities of a biological agents—anthrax and botulism toxin.

The Iraqi experience shows that biological warfare programs can exist and be hidden within legitimate facilities. Even with direct on-site visits the likelihood of discovery may be small unless the visitors know precisely where to go and are permitted entry. The existence of such programs can be enhanced with tight security and deception.

Recent events in the former Soviet Union concerning possible proliferation of nuclear materials also add greater uncertainty to the BW proliferation calculus. The potential impact and availability of scientists and technicians associated with the former Soviet offensive BW program should be considered. Aspiring proliferants may see them as lucrative recruiting targets. Unlike the complex technological requirements for nuclear weapons, one or two individuals with experience in BW agent production and weaponization could provide breakthrough assistance to a fledgling program.

The challenge for the United States is to develop a coherent and coordinated strategy to limit BW proliferation. If this fails, the United States must be prepared to deter, preempt, and defend against it. On the basis of the several incentives presented earlier, the likelihood of preventing further proliferation or rolling back existing BW-capable nations seems unlikely. The simple arithmetic of the number of countries suspected of offensive BW activities is at nine and likely to increase by the end of the century

Recommendations For The Future

How do we meet the BW threat in the twenty-first century? What policies will help solve the likely BW challenges? How should the United States counter-BW programs be prioritized and integrated?

Intelligence

Incomplete or absent intelligence about a suspected proliferant's BW program is a likely source of trouble. Not having specific information about the status of a BW program, or locations of production and storage, or methods of delivery, or the specific agents could result in an incomplete assessment. This could directly impact the development of United States strategy, policy, and capabilities to meet the threat.

A comprehensive anti-BW intelligence effort must collect information relating to the basic science, medical, and bioengineering capabilities of a potential proliferator. Short of having reliable human intelligence with direct access to an adversary's BW program, this type of information is required to assess the biological capability of that nation.

Intelligence collection and analysis are critical for future United States BW counter-proliferation efforts. Determining the intent to develop BW, locating suspect facilities, and assessing the nature of the offensive program are essential elements of the intelligence effort. The importance of specific BW agent intelligence for medical and detection capabilities deserves emphasis. Even if the intelligence community collects and validates this information, there may be a significant lag time, years or even decades, before safe, effective countermeasures can be developed and fielded.

Intelligence about the anticipated means of delivery and its doctrine of use is also important. This information allows development of US active defense capabilities to interdict and destroy delivery vehicles. Facility-related intelligence also allows for identification and targeting of production and related facilities for counteroffensive strikes.

The ability to identify BW-related organizations and proliferants is important to optimize the utility of export controls and the ability to interdict shipments of related equipment destined for suspected BW countries or organizations. The ability to focus limited diplomatic and economic resources to dissuade, pursue arms control, and bring international pressure on suspected proliferants is, therefore, also intelligence-dependent.

The likelihood that national technical means can identify any or all of the three key BW intelligence components—intent, location, and nature—is small. Dual-use facilities may not emit characteristic signatures, but still be capable of producing military significant quantities of biological agents. As alluded to earlier, the availability of human-source intelligence will be a critical element in providing information related to BW proliferation. Assessment of BW-related information requires trained personnel experienced in matters relating to biology, biotechnology, medicine, and agriculture. Balancing the technical component of the intelligence analysis process is the need to integrate the expertise and experience the intelligence system.

Because of the greater relative intelligence challenges and the myriad of related areas associated with BW versus nuclear proliferation, adequate resources must be applied to the problem. Increasing collection priorities should also necessitate a concomitant increase in the analysis resources devoted to the problem. There should be an assessment and if necessary a redistribution of assets to reconcile the disparity between the effort against nuclear proliferation and that of BW.

Policy and Strategy Development

Beyond the collection and assessment of intelligence, policy development and integration of the many functions are required to respond to BW proliferation. The domestic vulnerability to covert or clandestine acts of BW terror should be assessed. There must be executive-level interest and involvement to oversee the development of a crisis response system to domestic BW incidents. While the federal response to terrorist acts is well delineated, the time-sensitive health care demands created by an act of biological terror must be assessed. It is beyond the scope of the single agency identified in the Federal Emergency Response Plan, the Department of Health and Human Services, to mount the necessary reaction to deal with the health consequences and prevent unnecessary loss of life.

A historical example illustrates the scale of the effort required to respond to an act of BW terror in a major metropolitan area. In 1947, an American business man traveled to New York City from Mexico City. During his bus ride, he developed a fever, headache, and rash. Though ill upon his arrival in New York, he went sight-seeing. Over a period of several hours, he walked around the city and through a major department store. His illness, smallpox, progressed and he died nine days later. As a result of this single case, 12 other cases of smallpox and two deaths occurred. Because of smallpox's ability to be transmitted from person to person, this handful of cases was deemed so serious by public health officials that 6,350,000 persons in New York City alone were vaccinated in less than a month.[33]

Unlike 1947, Americans have not been routinely vaccinated against smallpox since 1980. A significant proportion of the US population is susceptible to this virus. The number of cases expected to occur as the result of a deliberate act could be in the thousands or tens of thousands. Even though the World Health Organization declared smallpox eradicated, North Korea has been identified as one possible country retaining cultures of this virus to use as a biological weapon.[34]

Even if a noncontagious agent were used, the public health consequences could be overwhelming. If several kilograms of an agent like anthrax were disseminated in New York City today, conservative estimates put the number deaths occurring in the first few days at 400,000.[35] Thousands of others would be at risk of dying within several days if proper antibiotics and vaccination were not started immediately. Millions of others would be fearful of being exposed and seek or demand medical care as well. Beyond the immediate health implications of such an act, the potential panic and civil unrest created would require an equally large response. Local law enforcement agencies would be overwhelmed and would need the assistance of state and federal agencies. The complete vulnerability of the United States if exposed to this type of terrorism would prompt other terrorists to attempt the same type of attack for extortion or additional terror impact.

Prior to a domestic incident such as this, a capable, practiced, and coordinated response mechanism must be in place. The Federal Emergency Management Agency (FEMA) provides this coordination function, but its actual familiarity and practice associated with biological terrorism is not known. The health-related support functions found in the Departments of Health and Human Services, Veterans Affairs, and Defense would have to be integrated into a single response plan.

Stockpiles of necessary antibiotics, immunoglobulins, and vaccines would have to be procured, maintained, and be readily available to administer within hours after recognizing an incident. An additional critical element of this response would be the management of information to allay fears and avoid unnecessary panic. The effort required to respond to a biological act of terror rivals that needed for an accidental or deliberate detonation of a nuclear device.

Arms Control

The BWC clearly represents the "lock which keeps the honest man honest." It serves a vital function by establishing an international norm against BW proliferation. Efforts should be made to provide both disincentives and incentives for current states to comply with the BW treaty. Nonsignatories should be leveraged to participate in the convention. Strengthening the BWC by enhancing transparency of biological activities is stated US government policy. Measures identified in the third review conference are being examined for use in a possible formal protocol. Like the warning to consumers—caveat emptor, subscribers to the convention must understand that complete verification is not just elusive but impossible.

During the original proposal of the BWC by the United Kingdom, the Soviet military strongly opposed any limitation on offensive BW. The Soviet Foreign Minister Andrei Gromyko felt that, for propaganda purposes, a prohibition on biological weapons would be useful. Without international controls, the Politburo and the Soviet military both endorsed the "toothless" convention in 1972.[36] The Soviet Union was one of the three depositories of this treaty. Recent disclosures by President Boris Yeltsin regarding former Soviet and Russian violations of the Biological Weapons Convention highlight the limits of that treaty.

The opportunity to strengthen the regime must also be tempered by events in Iraq since the Gulf War. Despite intrusive inspections, no concrete evidence of its offensive BW program has been uncovered, although Iraq has admitted to an R&D effort prior to the war. The continuation of the US sanctions is now solely dependent on Iraq's refusal to detail their BW program. Public statements by senior US intelligence and government officials indicates that Iraq retains its offensive BW capability. Indeed, there is a strong suspicion that Iraq may not have destroyed its BW agents as it has claimed.

A disturbing prospect for strengthening the BWC is that the 21 measures identified in the third review conference were utilized in Iraq including such actions as monitoring, sample collections, short-notice visits, record reviews, and interviewing of key personnel. These measures have yielded no conclusive evidence to date about the present Iraqi BW capability or its past program.

Diplomatic Measures and International Pressure

Current DOD counterproliferation policy emphasizes use of public diplomacy, positive and negative security assistance, and identifying the economic, political and military costs of proliferation. Denial of certain equipment or technologies used in BW is problematic. Export controls, interdiction, or disruption of supply networks will have limited impact given the dual-use legitimate nature of the biological materials and equipment.

Diplomatic efforts to prevent or control BW proliferation will have similar limits. States that are parties to the BWC, but are committed eventually to developing secret offensive biological weapons capabilities, can do serious BW research and development legally for a time within the current treaty framework. Nations who are not signatories can refuse entry and pursue offensive programs as well. Given the low likelihood of detecting violations, nonsignatories could take advantage of economic incentives or foreign assistance programs for joining the BWC, yet pursue clandestine offensive BW programs. There are reasonable indications that diplomacy alone can do little to prevent BW proliferation.

Nevertheless, diplomatic and economic pressure can serve a useful purpose in inhibiting a proliferant's activities by invoking sanctions, export controls, or publicly disclosing violations. One must realize that invoking these measures depends on timely and accurate intelligence. Diplomatic measures which try to control proliferation may, in the short run, delay a proliferant's efforts. In the long term, they may motivate determined proliferants to conceal or deceive their true intent and activities. Of course the perfect solution to proliferation is correcting the underlying reasons why nations choose to develop biological or nuclear weapons. Clearly, BW is a symptom of a deeper security need.

Countermeasures

Once proliferation occurs and an adversary attains an offensive BW capability, the focus changes to mitigating its perceived advantage and deterring its use. The range of counteroffensive capabilities and strategies must include deterrence, preemption, and destruction. During Desert Storm speculation occurred regarding the implicit use of nuclear weapons in response to BW attack. Both President George Bush and Secretary of Defense Richard Cheney publicly stated that any attack on US forces with chemical or biological weapons would be met with "massive retaliation." The accuracy and credibility of this policy option are subject to much debate.

Determining the culpable party after a covert or clandestine BW attack has occurred may be impossible. The circumstances following the bombing of Pan American Flight 103 highlight the potential difficulties of a forensic investigation following an act of terror. Before implicating Libyan intelligence operatives, both Iran and Syria and several terrorist groups were suspected. Finding a "smoking gun" and proving who is responsible for a future covert biological warfare attack could be difficult or impossible.

There will be times in the future when the US president may have to consider military preemptive strikes against a terrorist state or group to protect the United States and allied governments against BW attacks. Preemptive activities or anticipatory self-defense can range from efforts that occur during nonhostilities to open armed conflict. The recent interception and attempted interdiction of precursor CW materials bound for Iran from China (Yin He incident) present a recent example. The 1981 bombing of the Iraqi Orisak nuclear reactor by Israel is representative of using overt military force during instances short of open armed conflict. On the other hand, the US air campaign sought to destroy Saddam Hussein's WMD capabilities in the opening days of the Gulf War.

Destruction of the means to produce, process and deliver biological weapons is the final element of the counteroffensive strategy. There is a potential risk of collateral damage when striking BW-related facilities or delivery systems. Theoretically, a downwind hazard could occur if bulk storage of BW agents were struck when meteorological conditions were favorable to their dissemination. While no specific confirmed reports of collateral damage were documented during Desert Storm, there was one news report that implicated the occurrence of illness and death in Iraqi guards at an unidentified BW facility south of Baghdad after coalition bombing.[37]

Another partial answer to the BW threat would be to deploy theater ballistic missile defenses capable of intercepting enemy BW warheads while enroute to their targets. Ideally, interception and destruction would occur during early boost phase of the enemy missile launch to lower the risk of friendly casualties. Good ballistic missile defenses are not enough, since the United States and its allies must also respond to the challenge posed by future enemy cruise missiles equipped with biological weapon warheads. Of course, such missile defenses are defenses of the last resort. The greatest probability of minimizing collateral damage would be realized if special forces or other means of preemption allowed the United States or its allies to destroy the enemy BW capability prior to an adversary filling the weapon or launching the attack.

Biological Defense

The last element of the BW counterproliferation strategy is medical and nonmedical passive defensive measures. The importance and problems of BW detectors have already been identified. A priority effort exists to develop and field a BW detector which can provide stand-off warning and real-time detection of attack.

Another nonmedical defensive measure that deserves emphasis is the employment of collective protective systems. Hardened shelters and work areas for rear-echelon troops as well as filtered over-pressurized systems for combat vehicles, ships, and planes could minimize the effects of both chemical and biological weapons.

Medical measures to protect against biological threats include short-term and long-term methods of protecting US military forces. Even when detectors become available, medical measures will play an important role in both protection and treatment against BW attack.

Efforts should be focused first on developing, testing, and producing Food and Drug Administration-approved vaccines to immunize soldiers against the most likely BW threat agents. The availability of suitable vaccines and other medical products must remain a priority.

On 26 November 1993, Undersecretary of Defense William Perry signed the DOD immunization policy for BW threats. It establishes the requirement for both peacetime and contingency use of vaccines against validated BW threats. Each theater commander-in-chief (CINC) is required to determine the regional threat and provide the chairman of the Joint Chiefs with requirements. Integral in meeting the regional CINC's requirements is the development of a DOD-dedicated vaccine production infrastructure.

The liability concerns of the US pharmaceutical industry have affected development and production of public health vaccines. The controversy associated with the liability risk of immunizing children against Pertussis is illustrative. Congress mandated a government-subsidized fund to defray court awarded damages resulting from severe neurologic sequelae from the Pertussis vaccine. The rationale behind this effort was to protect the vaccine companies from large cash awards resulting from litigation, so to preserve their profitable production of important public health vaccines.

The commercial or public need for vaccines against biological warfare agents short of an act of terror is virtually zero. Yet, should a high-confidence warning of an attack on our population occur, substantial amounts of these products would be necessary to respond to minimize illness and death. The peacetime military need exists as result of the DOD immunization directive.

The ability and desire of the pharmaceutical industry to commit its facilities for dedicated vaccine development are questionable in light of profit and liability concerns. A US government vaccine facility has value for both BW and public health considerations. Such a facility should remain a high priority project in developing capability to respond to BW proliferation.

Besides the actual protective effect gained by BW vaccines, certain elements of deterrence can be garnered by minimizing the effect of an adversary's BW agent. Immunizing US forces and having the ability to protect others will minimize an enemy's ability to coerce the United States and its allies. They will also lessen the potential impact of a BW attack on the United States or its allies.

Therefore, the perceived or actual benefit derived by the BW proliferant will be lessened. Finally, vaccines and medical countermeasures can contribute the means to maintain the war-fighting capability of US military forces as well as providing for the survival of US citizens.

Conclusions

The proliferation of biological warfare weapons offers less developed nations a capability as lethal and potentially devastating as a nuclear device. The ease and relative low cost of BW production, coupled with spread of dual-use legitimate biotechnology, will facilitate and accelerate BW proliferation in the short-term and well into the twenty-first century.

Biological weapons can be employed in noncombat settings under the guise of natural events, during operations other than war, or can be used in open combat scenarios against all biological systems—man, animal, or plant. Deliberate dissemination of BW agents may be afforded possible denial by naturally occurring diseases and events. The low probability of detecting the development and production of terrorist and militarily significant quantities of BW agents lessens the effectiveness of diplomatic measures such as dissuasion, denial, and international pressure.

The limitations associated with treaty verification leave little optimism for the long-term effectiveness of the Biological Weapons Convention. Ascribing to the BWC may offer further potential of plausible denial if proliferants sought to use membership as a cover for their prohibited efforts.

Expectations for preventing BW proliferation must be grounded in reality. The likelihood of preventing or deterring a determined proliferant from obtaining biological weapons is relatively small. The outlook for the future of biological weapons proliferation is discouraging. "Brain drain" from the former Soviet Union may create volatile opportunities for breakthrough proliferants.

Future US policies against BW proliferation need to be based on integrated government policies and capabilities to deter, preempt and defend against this threat. No single element of the program is adequate to deal with the BW problem. Together, however, these elements can lower the risk and mitigate the potential impact of BW.

In addition, the problem of biological warfare cannot be narrowly focused on its ability to kill or render people ill. Biological warfare's potential to create significant economic loss and subsequent political instability with plausible denial exceeds any other known weapon. Germ warfare at the end of the twentieth and inception of the twenty-first century directly threatens the security of the United States and the achievement of a peaceful, prosperous, and stable post-cold war era.

Notes

1. Bradley Graham, "Battle Plans for a New Century," *The Washington Post*, 21 February 1995, A1.

2. Ibid., A4. See also, Andrew F. Krepinevich, "Cavalry to Computer, The Pattern of Military Revolutions," *The National Interest*, No. 37, Fall 1994, 30–42. These 10 revolutions in military affairs were identified as the infantry revolution, the artillery revolution, revolution of sail and shot, the fortress revolution, the Napoleonic revolution, land war revolution, naval revolution, mechanization revolution, aviation revolution, information revolution, and the nuclear revolution.

3. William H. McNeill, *Plagues and Peoples* (New York: Anchor Books, 1977).

4. Joseph D. Douglas and Neil C. Livingstone, *America the Vulnerable* (Lexington Mass. Lexington Books, 1987) ,2.

5. "World AIDS Day December 1 1993" *Morbidity and Mortality Weekly Report* (Atlanta: Centers for Disease Control, 19 November 1993), 1.

6. Congress, Office of Technology Assessment, *Proliferation of Weapons of Mass Destruction: Assessing the Risks* (Washington, D.C.: GPO, 1993), 3.

7. Stanley L. Weiner and John Barret, *Trauma Management for Civilian and Military Physicians* (Philadelphia: W.B. Saunders, 1986), 519.

8. General Accounting Office, *Chemical & Biological Defense: U.S. Forces Are Not Adequately Equipped to Detect All Threats* (Washington, D.C.: USGPO, 1993) 1.

9. Ibid., 3.

10. Arthur Friedlander, et al., *Post-exposure Prophylaxis against Experimental Inhalational Anthrax* (Frederick, M.D.: Army Medical Research Institute of Infectious Diseases, 1990).

11. Jonathan Lasch, "Human Monoclonal Antibodies for Biological Warfare Defense," (La Jolla California: The Scripps Research Institute, 19 November 1993), 1.

12. George F. Brooks, et. al., *Medical Microbiology*, (Norwalk, N.C.: Appleton and Lange, 1991), 162–163.

13. Friedlander.

14. Robert Harris and Jeremy Paxman, *A Higher Form of Killing* (New York: Noonday Press, 1982), 76. Also Defense Intelligence Agency, *Soviet Biological Warfare Threat* (Washington, D.C.: USGPO, 1986), 2. Another good source is Michael R. Gordon, "CIA Foresees Iraq Biological-Weapon Capability in Early '91," *New York Times*, 20 September 1990, 1.

15. David L. Huxsoll, "The U.S. Biological Defense Research Program," in *Biological Weapons: Weapons of the Future?* ed. Brad Roberts (Washington, D.C.: The Center for Strategic & International Studies, 1993), 60.

16. Robert Gates, *McNeil and Lehrer News Hour*, 20 January 1992.

17. General Accounting Office, *Biological Warfare: Better Controls in DOD's Research Could Prevent Unneeded Expenditures* (Washington, D.C.: USGPO, 1990), 5.

18. Office of Technology Assessment, 39.

19. Army, *U.S. Army Activity in the U.S. Biological Warfare Programs*, vol. 2 (Washington, D.C.: USGPO, 1977), D 2, 3.

20. Ibid., 51.

21. Office of Technology Assessment, 39. See also, Robert H. Kupperman and David M. Smith, "Coping With Biological Terrorism," in Roberts, 41.

22. Office of Technology Assessment, 41.

23. Ibid., 54.

24. Ibid., 8.

25. Ibid.

26. R. Jeffrey Smith, "Iraq Had Program for Germ Warfare," *The Washington Post*, 6 July 1995, 1.

27. George Friel, quoted in "Chem-Bio Defense Agency Will Tackle Last Major Threat to a Deployed Force," *Armed Forces Journal*, 1992, 10.

28. General Accounting Office, 1.

29. Office of Technology Assessment, 11.

30. Ibid., 39.

31. Ibid., 80.

32. Ibid., 38.

33. Theodore Rosebury, *Peace or Pestilence: Biological Warfare and How to Avoid It* (New York: McGraw-Hill Book Co., 1949), 60.

34. John J. Fialka, "CIA Says North Korea Appears Active in Biological, Nuclear Arms," *The Wall Street Journal*, 25 February 1993, 16.

35. Kupperman and Smith, 42.

36. Arkady N. Shevchenko, *Breaking With Moscow* (New York: Alfred A. Knopf, 1985), 174.

37. "Biological Plant Bombed," London: *Reuters Wire Service*, 3 February 1991.

Chapter 10

Biological Weapons for Waging Economic Warfare

Lt Col Robert P. Kadlec, USAF

The final decade of the twentieth century has positioned the world at the threshold of tremendous opportunity. The collapse of the Soviet Union has dissolved the bipolar world and created the opening to forge a new international security environment. The preeminence of politico-military competition is slowly giving way to politico-economic competition. As Shintaro Ishihara predicts, "The twenty-first century will be a century of economic warfare."[1]

While military power remains important, its context and type are changing. The focus of many developing nations is to seek weapons of mass destruction (WMD)—nuclear, biological, and chemical weapons—to meet regional security concerns. The parallel emergence of economic competition and its likely accompanying conflicts with the proliferation of WMD raises the possibility of a new form of warfare. This includes the development and use of biological warfare (BW) against economic targets.

Using BW to attack livestock, crops, or ecosystems offers an adversary the means to wage a potentially subtle yet devastating form of warfare, one which would impact the political, social, and economic sectors of a society and potentially of national survival itself.

Agriculture

For both developed and developing nations, nonfuel commodities present an important source of national security and prosperity. In the United States alone, the agricultural sector is an $800 billion industry. Besides providing for the nourishment of the US population and a significant portion of the world, agriculture generated approximately $67 billion in export revenues in 1991. This revenue represents approximately 15 percent of the total US exports for that particular year.[2] Agricultural exports have been an important source for redressing the US trade deficit. Moreover, agriculture is now one of but a handful of sectors that generates a trade surplus for the US. In 1992 it created an estimated $18-billion surplus.[3]

Lesser developed and developing nations and other nations whose economies are in transition have significant agricultural sectors that provide important contributions of food and revenue to their economies. This observation is especially true of nonoil producing nations. Yet, even with productive agricultural systems, most if not all nations in the world are food importers.

Trends in agricultural systems, particularly food production, indicate that fewer numbers of people and hectares are involved in agricultural production. In developed market economies, the percentage of the economically active population in agriculture declined by 31.2 percent from 1980 to 1992.[4] A similar, yet not as dramatic, decline was noted in developing countries, where the numbers of people involved in agriculture declined by 11.3 percent during the same period.[5] Despite that decline, the overall agricultural productivity in both the developed and developing worlds increased by 45.3 percent and 25.2 percent respectively.[6]

This increase in productivity has resulted from the spread of modern farming technology, high-yield crop varieties, and potent fertilizers and pesticides. The goal of many developing and developed nations is to become self-sufficient in food and other agricultural products.[7] Competition has become intense.

Efforts to remove trade-distorting domestic subsidies and limits to market access to agriculture were objectives of the Uruguay Round of the General Agreement on Trade and Tariffs. Market access-limitation policies essentially maintain domestic prices above world prices and isolate domestic producers from competition and the volatility of the world market.[8] While included on the Uruguay Round's agenda, tremendous resistance was encountered from several important nations. The United States wanted to protect dairy products, sugar, cotton, and peanuts. Japan wanted to prevent rice imports. Despite efforts to settle differences on issues of market access, internal supports, and export competition, agreement on many items was not reached.

Biotechnology

Part of the economic revolution in the world today is the explosion of biotechnology. Biotechnology has been a significant reason why agricultural systems are much more productive. As alluded to earlier, the development of higher-yield crops results partly from genetic recombinant engineering, which takes genes coded for greater productivity and resistance to disease and drought and inserts them into a particular species of crop.

Besides enhancing the productivity and heartiness of food or cash crops, methods of biological control are increasingly relied upon to provide an environment-friendly means of controlling economically significant pests and diseases. Bacillus thuringiensis (B.t.). is a well-known example of a naturally occurring sporulated bacteria which effectively controls caterpillars, particularly tomato worms.

A variant of B.t., called B.t. israelensis or B.t.i., has shown its effectiveness in controlling malaria-bearing mosquitoes and blackflies which carry the parasite that causes river blindness.[9] Efforts are now under way to insert the gene from B.t. into such plants as cotton. Initial research indicates that this procedure enables cotton plants to resist the boll weevil (anthonomus grandis). This particular pest caused an estimated $50-billion loss in US cotton revenues from 1909 to 1949.[10]

In California's Imperial Valley the pink bollworm caterpillar has caused the amount of land planted with cotton to drop from 140,000 acres to only 7,000 during the past 17 years.[11] Today US cotton farmers spend $500 million on pesticides.

Nature of the Biological Warfare Threat

Harmful bacteria, viruses, rickettsia, or toxins that incapacitate or kill humans, animals, or plants have an unsettling value in waging economic warfare. In 1925 Winston Churchill envisioned a context for BW when he wrote about "pestilences methodically prepared and deliberately launched upon man and beast . . . Blight to destroy crops, Anthrax to slay horses and cattle . . . ."[12] This discussion narrows the definition of BW to consider only its utility against such economic targets as animals and plants.

Historical Context and Evolution

Investigators have argued that German agents intentionally infected horses and cattle with anthrax and glanders before they were shipped from the United States to Europe during World War I.[13] During World War II, the United States, fearful of perceived efforts by both Japan and Germany to develop BW, engaged in a large and ambitious retaliatory offensive and defensive BW research and development effort. While never fielding or using a BW weapon, they did develop several BW agents, including rinderpest, glanders, wheat rust, rye rust, and rice blast to use against animals and plants.

Anecdotal reports suggest US officials had considered using rice blast agents to destroy Japan's rice crop during the closing months of the war to force its surrender. The realization that the United States would have to supply food to Japan once the war ended and the availability of the atomic bomb, dissuaded US officials from pursuing this option.

In 1972, an international treaty, the Biological Warfare Convention, specifically prohibited the research, development, production, or use of biological agents for offensive use. While 162 countries have signed this treaty, no verification means are available to ensure compliance. Reportedly, up to 20 nations are suspected of pursuing offensive BW capabilities. Significant on the list are Russia, China, Iran, Iraq, Syria, Israel, North Korea, and Taiwan.[14] No specific mention is made of any suspect nation seeking development of anti-animal/ anti-crop agents. Note that the United States during its offensive program first developed and fielded an anti-crop bomb. The United States discontinued its pursuit of several anti-agricultural agents in the mid-1950s since they lacked military utility.

Biological Weapons: Cost-effective WMD

Compared to other mass destruction weapons, biological weapons are cheap. A recent Office of Technology Assessment (OTA) report places the cost of a BW large arsenal as low as $10 million.

This estimated cost stands in stark contrast to a low-end estimate of $200 million for developing a single nuclear weapon. The high-end cost estimate for a nuclear weapons could be 10 to 50 times higher.[15] Not only is BW more affordable, but militarily significant quantities of BW agents (kilograms) in legitimate biological laboratories make BW production easy to accomplish and conceal. Any nation with a moderately sophisticated pharmaceutical industry can do so.

Nature at Work: Whiteflies and Plausibility

Biological economic warfare likely would involve the intentional dispersion of a harmful agent or pest against a high-value cash crop or food source. The US Department of Agriculture recently identified 53 animal diseases which are nonindigenous or foreign, which, if introduced into this nation, would adversely impact the livestock industry.[16] Recent naturally occurring events highlight this potential.

The Imperial Valley produces a large variety of food and produce. In the summer of 1991, an infestation by the sweet potato fly or whitefly destroyed much of the crops in this area and caused a $300-million loss. A related but different strain of whitefly caused $100 million in losses in southeastern California in 1981. The US agricultural system is a $800-billion industry. The Imperial Valley infestation represents a natural event where a harmful agent (whitefly) encountered a susceptible host (crops) in a conducive environment (the Imperial Valley). The investigation of this natural outbreak, however, reveals just how a deliberate act of BW economic warfare could be engineered.[17]

The poinsettia strain of the whitefly is not found naturally in California. In the circumstance of this outbreak, the whitefly could have accompanied a shipment of poinsettia plants from Florida. While the exact place the poinsettia strain originate remains a mystery, other similar strains originate in Russia, mainland Asia, and Africa.

In its natural habitat, the whitefly has a certain homeostatic existence. Balanced between natural conditions, competitors, pathogens, and predators, the impact it has on the environment is usually limited. When this fly or any other pest is placed in an environment where natural controls are missing, uncontrolled insect breeding may cause subsequent crop destruction. In the Imperial Valley circumstance, however, the culpable insect represented more than simply a pest translocated to new fertile fields. This particular type of whitefly was a distinct new strain. Its biological characteristics made it an effective agent of destruction. Its appetite was voracious. Unlike other known strains of whiteflies, this one consumed many times its body weight in vegetation and dined on a great variety of plants. Second, it had an unusual resistance to chemical and naturally occurring pesticides. DNA analysis of the genetic makeup showed a unique strain of this particular insect. Finally, besides its direct effects, the whitefly carried other harmful agents like fungus. Thus, it also inflicted disease on already weakened plants.

Naturally occurring genetic events of mutation and selection reasonably explain this occurrence. It is also possible that such insects could be bred for nefarious purposes. In the context of a deliberate act of BW, a nation could select from several native occurring or endemic pests. Selective management and breeding could develop a "super" pest. The selection of this pest could be highly specific for a particular crop that an economic competitor or regional adversary relies on for economic prosperity or national survival. To provide better cover for a clandestine or covert BW attack, pests endemic to the target nation could be similarly obtained and could enhance its resistance through such laboratory manipulation as nonindigenous pesticide exposure. Infiltrating and disseminating perpetrator insects is then dependent on the mode of transportation and the level of plausible denial desired.

United States Vulnerabilities

The threat of this type of insect-borne BW attack on the United States remains theoretical. A recent OTA report on the

United States addressed the threat from harmful nonindigenous species (NIS). The report indicated that the intentional (noncriminal) or unintentional importation of plants, animals, or microbes has major current and future economic consequences for US agriculture, forestry, fisheries, water use, utilities, and natural areas.

Importation of harmful nonindigenous species costs the United States billions of dollars annually.[18] From 1906 to 1991, 79 NIS caused documented losses of $97 billion (Table 3). This table detailed only a small percentage of the large number of economically and environmentally costly agents, so their true impact is not known.

Table 3

Estimated Cumulative Losses to the US from Selected Nonindigenous Species, 1906–1991

Category	Species Analyzed (number)	Cumulative Losses ($ millions, 1991)	Species Not Analyzed
Plants[a]*	15	603	—
Terrestrial Vertebrates	6	225	>39
Insects	43	92,658	>330
Fish	3	467	>30
Aquatic Invertebrates	3	1,207	>35
Plant Pathogens	5	867	>44
Other	4	917	—

[a]Excludes most agricultural weeds.

Source: M. Cochran, "Non-Indigenous Species in the U.S. Economic Consequences," prepared for Office of Technology Assessment, March 1992.

US losses between 1987 and 1989 to the Russian wheat aphid (diurahis noxia), for example, exceeded $600 million.[19] The Mediterranean fruit fly caused $897 million in damage and lost revenue. Each year $7.4 billion is spent on pesticide applications, with a significant amount spent on controlling NIS insects. Nonindigenous weeds, with both direct and indirect effects, caused a loss of somewhere between $3.6 and $5.4 billion per year. If herbicides were not used to control them, weed loss would hover around $19 billion yearly.

Another recent example cited in the OTA report described how NIS gain entry into the US. The Asian tiger fly (anopheles albopictus) mosquito does not naturally live in the US. It is normally found in Southeast Asia where it is the vector or carrier for the human diseases dengue and malaria.

In 1985 a freighter carrying containers of old tires imported this mosquito into the United States via the Port of New Orleans. This mosquito is an aggressive human biter and a prolific breeder. Because of its behavior, the Asian tiger fly poses a greater risk of endemic or naturally occurring mosquito-borne disease transmission. With no geographic barriers, the tiger fly has spread to 22 states and is creating a public health concern because of the increased occurrence of Western and Eastern equine encephalitis and the reemergence of dengue fever in the United States.

The impact and magnitude of the tiger fly will not result in billions or millions of dollars of lost productivity or tens of thousands or thousands of deaths. Clearly, the United States has a well-established public health system with surveillance, rapid identification, and management if an epidemic or outbreak occurs.

Nations of the third world, however, are not as fortunate and do not have an existing infrastructure nor adequate resources to mitigate the impact of similar events. This shortcoming was recognized in an epidemic of yellow fever in Nigeria in 1991. A shipment of used tires from Asia was implicated in the introduction of this insect.[20] Similar modes of NIS infiltration have been described as a result of airline travel and flushing ballast tanks on ships.

A contemporary theoretical example of a third world BW economic scenario is represented by an actual situation in Malaysia, the world's third largest producer of rubber behind Thailand and Indonesia. In 1991, it exported $971.9 million of natural rubber. Along with other Southeast Asia nations, Malaysia is trying to keep the South American leaf blight (microcyclus ulei) from affecting its rubber tree industry. This fungus was first detected in Brazil at the turn of the century and infects the stems of young trees and leaves and significantly decreases the output of sap.

No known cure for microcyclus exists. This blight is the main reason a viable rubber industry no longer exists in South America. The immediacy of airline travel, especially directly from Brazil to Malaysia, makes possible the unintentional entry of this fungus. Estimates by Malaysian Agricultural Department officials predict that should this fungus enter into its country's rubber trees, it would decimate the trees within two years. Fighting off this fungus is considered vital to sustain Malaysia's economic boom at its projected 7–8 percent growth rate.[21]

In the spring of 1993, an outbreak of hoof-and-mouth disease occurred in Italy. This contagious virus led to the destruction of 4,000 head of cattle. After authorities linked the source of this outbreak to a live cattle shipment from the former Yugoslavia, the European Community (EC) sparked a "cow war" when they banned bovine products from all 18 countries of Eastern Europe and the former Soviet Union. Bovine meat exports from eight East Europe countries (Bulgaria, Czechoslovakia, East Germany, Hungary, Poland, Romania, and a portion of the former Soviet Union) totaled $103 million in 1991. Poland, Czechoslovakia, Hungary, and Bulgaria reacted by banning the EC's own meat exports. Eastern European officials say EC markets remain almost as closed to them in 1993 as to their former communist regimes.[22]

While the governments of Eastern European countries have large agricultural infrastructures with the potential for large exports, they lack money and do not conform to the EU's common agricultural policy. The cow wars restrict their access to agricultural markets and economic capital which could finance greater economic and political reforms.

The reappearance of the screwworm along the Mexico-Texas border has worried US cattlemen. Its return resulted from importation of an infected herd from Central America in November 1992. Infection of US cattle would result in "severe economic losses."[23] This pest eats the flesh of cows, destroying their hides, and kills newborn calves. During the past four decades, over $400 million has been spent on US-Mexican screwworm eradication programs. A major concern created by lowering trade barriers during the North American Free Trade Agreement negotiations was how this treaty would facilitate the spread of agricultural pests like the screwworm.

Illustrative Scenarios

Thus, naturally occurring events where the agent, the susceptible host, and the environment converge can result in disease, economic loss, and national or international repercussions. No evidence indicates that any nation or group willfully caused the events cited. However, we may not be so lucky in the twenty-first century. It is all too possible to construct a scenario which would offer plausible denial and possible gain to a potential adversary.

Scenario One: Corn Futures

The US Department of Agriculture estimates that the 1994 corn harvest would plunge 31.4 percent from record summer rains and flooding. It was expected the cost of a bushel of corn would soar to three dollars. The February 1994 price of corn, the nation's number one crop, was the highest in a decade.[24] The resultant increase in cost increased operating expenses of companies "that handle[d] everything from corn-based ethanol fuel to livestock."[25]

"We are vulnerable in 1994 . . . right on the edge," said Keith Collins, acting assistant agriculture secretary for economics. The fall 1993 harvest was too small to supply both domestic processors and exporters. US stockpiles are expected to be at their lowest levels since the food scare of the mid-1970s. A slight acceleration in food inflation was expected in 1994. Food inflation in 1994 was estimated at between 3.3 percent and 3.5 percent, up from 2.2 percent in 1993 and 1.2 percent in 1992. This is the first time in about four years whereas food prices exceeded the general inflation rate.[26]

A corn crop short of the 8.4-billion-bushel estimate would signal a 4 percent food inflation rate in 1995.[27] Some additional disasters such as a drought or loss of corn to blight, "would do considerable economic damage to world food supplies."[28] A 7.5-billion-bushel corn crop would push prices to four dollars a bushel. Such a surge would push inhibited exports and make hogs and cattle too expensive for many farmers to feed, eventually driving up meat prices.[29]

How could someone use biological agents to conduct economic warfare by ruining a competitor's crop or product? Three more scenarios involving corn, wine, and cotton attacks can illustrate the potential BW threat.

Scenario Two: Corn Terrorism

A second scenario might go as follows. China is the world's second largest corn exporter.[30] Recognizing the vulnerable situation of the United States, China plans an act of agricultural terrorism. Selecting a corn seed blight, fusarium graminearum, which grows well at cool temperatures and in wet soil, they clandestinely spray this hearty spore over the US Midwest from commercial airliners flying the polar route to Chicago and Saint Louis. They disseminate the spore in winter and early spring, and the blight is present in the soil when spring planting occurs.

The United States, despite eliminating the corn set-aside requirements and planting more than 80 million acres of corn, suffers from a crop disaster. This unexpected Chinese-induced corn seed blight decimates the US corn crop. The fall harvest is a full 30 percent below expected levels. The United States then imports corn for the first time in its history to meet domestic needs. Food prices rise sharply and cause higher-than-expected food prices and inflation. China gains significant corn market share and tens of billions dollars of additional profits from their crop.

Scenario Three: That's a "Lousy" Wine

The grape louse (phylloxera vastratrix) is ravaging the vineyards of Napa and Sonoma counties in Northern California. It is estimated that 60–70 percent of the 68,000 vine acres are being destroyed. The louse does not affect the fruit of the vine but attacks the roots, which slowly kills the plant. It is difficult to detect, and once in place, it is a prolific breeder. The louse is carried by wind, water, and mud and, once discovered, is likely to have spread already to other areas.[31]

Infected vines may be treated by removing them and replanting them with louse-resistant plants. The estimated cost to the Napa-Sonoma wine industry will be about $1 billion or more over the next few years. The expected impact will be heavier on the more expensive Northern California wines, causing the disappearance of some labels.

A hypothetical scenario could consist of a group of disgruntled European winemakers who are jealous at the superior quality of Northern Californian wines and desirous of recapturing the wine market. Traveling as tourists, they slip into the United States with tins of paté which conceal millions of the offending louse. Traveling through the California wine country, they disperse their deadly cargo.

Scenario Four: Sabotaging Pakistan's Cotton Crop

Pakistan is the world's third-largest producer of cotton behind the United States and China. In 1991 the Pakistanis exported $420.4 million of raw cotton, some 36 percent of its primary commodity exports (excluding fuels) in 1991.[32] If textiles, yarn, and other by-products are included, almost 60 percent of Pakistan's exports depend on cotton.[33] Due to an attack caused by an insect, the 1993 harvest will fall 15 percent below expected levels.

This crop loss will affect the country's overall economic performance. Pakistan will produce only 10 million or so bales of the 12 million bales targeted. In 1993, Ahmed Muktar (the minister of state for commerce) said, "This definitely would be detrimental to our economy, because the surplus ... would have added to our meager foreign exchange reserves."[34] The immediate economic impact of the crop loss, however, may have longer-lasting effects. Cotton farmers, fearful of experiencing a similar disaster next year, are considering planting something else. Rice, wheat, and sugarcane, which are significantly less profitable (cotton is 43 times more profitable than all other crops), appear more attractive and safer than cotton.[35] "Restoring the confidence of farmers, who doubt their ability to generate profits from cotton, could become one of Pakistan's toughest challenges."[36]

The open hostility between Pakistan and India is not hypothetical. They are competing against each other in an arms race involving nuclear and conventional weapons. The heavy reliance of Pakistan on a single export cash crop is not unusual in the developing world. The geographic proximity of

Pakistan to its principal adversary allows a fairly easy route of infiltration and introduction of a pest against a high-value target like its cotton crop. The ability to inflict economic loss on Pakistan has repercussions that affect the private and commercial sectors, the military, and the society.

Implications of Economic Biological Warfare

The current US focus on BW limits consideration to the human effects of such agents as anthrax, plague, and smallpox. Little or no effort seems to be devoted in assessing the vulnerabilities of the United States or any other nation's agricultural or ecological infrastructures to BW attack. If the focus of international and regional competition is transitioning to economic power, it is prudent to assess the potential impact of this form of economic warfare, develop a comprehensive surveillance or monitoring system, and prepare countermeasures.

Developed countries with adequate economic reserves, agricultural diversity, and the means to mitigate such occurrences would be relatively resistant to such attack. Even developed countries, however, could experience significant economic losses and political and national security repercussions if other intervening events could make certain target commodities more vulnerable or magnify the impact of BW use.

Lesser developed or developing nations are in a much more precarious position. If the target commodity was a principal cash crop or food source, using BW may inflict a grave blow to that nation's economy or society and possibly result in some political impact. History has recorded the chaos and instability created by such natural catastrophes as famines and epidemics. Using BW in this fashion would have applications to waging low-intensity warfare with strategic outcomes.

Addressing the Problem

Recent public debate about the appropriateness of the US intelligence community's collecting economically relevant intelligence did not mention the impact of BW proliferation on economies. Any open-source discussion about proliferation of biological weapons does not address its utility in waging economic warfare.

As in other fcrms of weapons proliferations, however, intelligence remains key. The foundation of intelligence assessment related to BW directed against economic targets is based on human intelligence. Collection activities must be focused on research and development efforts of adversarial nations in areas relating to endemic and nonendemic nonhuman diseases and pests. The means to assess any information collected requires a truly multidisciplinary effort involving veterinarians, ecologists, horticulturists, botanists, entomologists, and intelligence analysts.

While there is a requirement to assess what potential adversaries are doing in these areas, vulnerability and potential impact data must be collected on US systems. This effort requires a coordinated interagency process involving the Departments of Agriculture, Interior, Commerce, and Treasury as well as the Environmental Protection Agency. Besides sensitizing the federal government to the potential problem, state and local governments should support this effort. State and local governments should be educated on the importance of reporting nonhuman outbreaks of disease or pests with economic significance.

Local and federal agencies should investigate reports of nonhuman outbreaks which occur in defined high-value commodities, involve potential BW or nonendemic agents, or inflict a certain threshold economic loss.

Similarly, some existing integrated governmental mechanism must be mobilized quickly to contain and mitigate the impact of a BW attack. The Federal Emergency Response Plan contains at least a theoretical structure to begin to address this problem.

The real and hypothetical examples cited highlight the opportunity offered by BW as a means to attack the agricultural infrastructure of an adversary. The existence of naturally occurring or endemic agricultural pests or diseases and outbreaks as described permit an adversary to use BW with plausible denial.

The impact of such events would go beyond simply affecting a nation's economy to potentially affecting its national security. The United States gave up its antiagricultural biological weapons long before it unilaterally renounced the use and development of biological warfare in 1969.

The present concerns about the proliferation of weapons of mass destruction appropriately recognize the threat posed by BW against our military and citizenry. The question is whether our government is aware of, or prepared to respond to, acts of BW? Is our intelligence community sensitized or tracking proliferant's efforts to develop antiagriculture BW? Is there a mechanism whereby federal, state, and local agencies report and respond to acts affecting valuable economic resources or involving suspicious or nonendemic agents?

In the post-cold war era and as we enter the twenty-first century, the economy determines superpower status. The threat posed by biological weapons deserves prudent consideration.

Notes

1. Shintaro Ishihara cited in C. Fred Bergsten, "Japan and the United States in the New World Economy," in Theodore Rueter, ed., *The United States in the World Political Economy*, (New York: McGraw-Hill, 1994), 175.
2. *UNCTAD Commodity Yearbook*, (New York: United Nations Publication, 1993), 2.
3. *World Economic Survey 1993: Current Trends & Policies in the World Economy* (New York: United Nations Publications, 1993), 72.
4. Ibid., 155.
5. Ibid.
6. Ibid.
7. Ibid., 72.
8. Ibid., 71.
9. Thomas Y. Canby, "Bacteria: Teaching Old Bugs New Tricks," *National Geographic*, 184, no.2 (August 1993); 53.
10. U.S. Congress Office of Technology Assessment, *Harmful Non-indigenous Species in the United States* (Washington, D.C.: US Government Printing Office, 1993), 65.
11. Jon Thompson, "King of Fibers," *National Geographic*, 185, no. 6. (June 1994): 80.
12. Robert Harris and Jeremy Paxman, *A Higher Form of Killing* (New York: Noonday Press, 1982), 70.
13. Rexmond C. Cochrane, "Biological Warfare Research in the United States," in *History of the Chemical Warfare Service in World War II (1 July 1940–15 August 1945)* (Washington DC: US Army, 1946), 3.

14. Office of Technology Assessment, *Proliferation of Weapons of Mass Destruction* (Washington, D.C.: Government Printing Office, 1993).
15. Ibid., 11.
16. Office of Technology Assessment, *Harmful Non-indigenous Species in the United States*, 301.
17. Raymond J. Gill, "The Sweet Potato Whitefly Problem in Southern California," *California Department of Food & Agriculture Report* (Sacramento, Calif.: 1991).
18. *Harmful Non-indigenous Species*, 5.
19. Ibid., 65.
20. "Aedes albopictus Introduction into Continental Africa," *Morbidity and Mortality Weekly Report* (Atlanta: Center for Disease Control, 6 December 1991), 836.
21. "Malaysia Rubber Industry Fears Deadly Latin Fungus," *Baltimore Sun*, 18 April 1993.
22. Eric Bourne, "Europe's 'Cow War' Shows Persistence of Cold War Divides," *Christian Science Monitor*, 20 April 1993. See also, "Epidemic Hits Italian Livestock," *Wall Street Journal*, 16 March 1993.
23. David Clark Scott, "Pest Infestation Spurs Doubts Over Standards in Mexico," *Christian Science Monitor*, 16 January 1993.
24. "The Outlook: As Food Prices Rise 1994 Crops Hold Key," *Wall Street Journal*, 7 March 1994, 1.
25. Scott Kilman, "Corn Prices Are Expected to Soar After Forecast of 31.4% Plunge in US Harvest From Last Year," *Wall Street Journal*, 9 November 1993.
26. "The Outlook: As Food Prices Rise 1994 Crops Hold Key," 1.
27. Ibid.
28. Ibid.
29. Ibid.
30. *World Economic Survey 1993*.
31. Kathleen Sharp, "The Louse and the Grape," *American Way*, 15 September 1993 44.
32. *UNCTAD Commodity Yearbook*, 16.
33. Farhan Bokhari, "Cotton Farmers Face Big Setback," *Christian Science Monitor*, 29 December 1993, 8.
34. Ibid.
35. Ibid.
36. Ibid.

Chapter 11

On Twenty-first Century Warfare

Dr. Lawrence E. Grinter and Dr. Barry R. Schneider

Shortly after he became secretary of defense, William S. Cohen toured US forces in Korea. The briefings he received on the North Korean chemical and biological warfare threat caused him to immediately order a billion-dollar increase in US defense spending on counterproliferation programs: $732 million for passive defenses, $146 million for counterforce programs, $87 million for special forces improvements, and $36 million for active defense additions. It is a fact of life that even poor states such as the Democratic People's Republic of Korea, if determined, can develop revolutionary weapons that might offset the conventional firepower of the world's sole remaining military superpower.

In order to plan for the battlefield of the future, the chairman of the Joint Chiefs of Staff recently released a report called *Joint Vision 2010*. This report offers a broad framework for understanding joint warfare in the future. It also directs US armed forces acquisition programs in obtaining the capabilities to run military operations using precision engagement, dominant maneuvers, and focused logistics while providing full dimensional protection of forces and assets in combat. *JV2010* is a reflection of the basic belief that a revolution in military affairs is occurring, and it assumes that new capabilities are needed to cope with that RMA.

There have been perhaps a dozen revolutions in military affairs since the fourteenth century.[1] One recent one began when the United States detonated its first atomic bomb in the early 1945 Trinity test. The Hiroshima and Nagasaki detonations in August 1945 helped end World War II. In that same war, the Japanese experimented with germ warfare by spreading bubonic plague agents on Chinese population centers via bombing missions. Also in WWII, the Germans manufactured—but did not use—Sarin and Tabun nerve gases.

These nuclear, biological, and chemical armaments now form a new "trinity" of weapons of mass destruction that threaten to make twenty-first-century warfare more costly than anything seen before.

In 1998 the world has seven acknowledged nuclear weapons states: the United States, Russia, the United Kingdom, France, China, India, and Pakistan. (India and Pakistan do not yet admit their nuclear test explosions are for weapons.) There is at least one undeclared nuclear weapons state, Israel. Beyond this, there are the near-nuclear states such as North Korea, Iran, and Iraq. If biological and chemical weapons arsenals are added to the nuclear club, it is estimated that between 20 and 30 states possess one or more of the NBC and missile weapons. Some of these are hostile radical regimes—rogue states that threaten their neighbors with intervention and/or state-sponsored terrorism.

Another revolution in military affairs, at least from an American perspective, has been the spread of WMD to such NBC-Arming Sponsors of Terrorism and Intervention (NASTI). This will cause considerable change in how the United States and its allies will have to fight, train, equip, and supply their forces opposing such foes in future major regional conflicts.

In a conflict against such a heavily armed opponent, will the United States and its allies be forced toward more dispersed forces, greater mobility, outranging of the opponent in disengaged combat, risky reliance on escalation dominance to deter adversary use of WMD, and/or improved active and passive defenses? Or can we follow an enhanced version of Desert Storm where, relying on escalation dominance to deter enemy intrawar resort to WMD, we emphasize parallel warfare, hyperwar, information warfare, dominating maneuver, precision targeting, and/or space technologies to beat the enemy military and secure our objectives?

Which set of technologies will be used in twenty-first-century wars? Which set of strategies best fits those technologies? Will the technologies of a past RMA, in the hands of our enemies, neutralize or preclude the utility of the newer technologies of a more modern RMA? Translation: will enemy use of NBC and missile systems influence the terms of battle so as to marginalize our ability to prosecute information warfare, space war, precision warfare, and dominating maneuver? Or will the fear of allied retaliation keep the adversary from initiating WMD strikes against allied forces, ports, air bases, and cities?

High-altitude nuclear bursts and the resultant electromagnetic pulse (EMP) might render most allied space assets inert. EMP could burn out the circuitry of most allied radio systems, computers, transistors, and power grids in the region of combat, rendering many of the allies' high-tech assets harmless.

Adversaries could also mount NBC weapons on mobile missile launchers, camouflaged, constantly moved, and hidden from sight and easy detection. This would create a targeting nightmare similar to that the allies faced in the Gulf War against Iraq's Scud missile launchers.

Countermeasures taken to blind US and allied space assets may rob war-fighting commanders in chief and their staffs of the information needed to target enemy forces, especially their highest-valued mobile military assets. After all, missiles and smart bombs still need correct coordinates to carry out precision attacks. Further, if the enemy is not blinded effectively, the dominating maneuvers of future "left hooks" may be rudely interrupted by catastrophic encounters with nuclear attacks and anthrax barrages or a battlefield engulfed in clouds of poison gases.

Would the NBC and missile revolution in military affairs trump the Warden RMA of parallel war/hyperwar? Would it overcome the RMA identified by the analysts from Science Applications International Corporation (SAIC) which emphasizes the contributions that may be made by changes in a combination of precision war, space war, information war, and dominating maneuver technologies, organizations, and strategies? Or will theater missile defenses emerge that can neutralize an adversary armed with 20 to 40 NBC warheads mounted on ballistic and cruise missiles?

Alternatively, even if effective ballistic and cruise missile defenses are not developed, will the allies' possession of overwhelming nuclear weapons preponderance be enough to deter rogue states from using their more limited WMDs once war has begun? Will deterrence still suffice to keep the peace even if effective missile and air defenses are not available? Clearly, US strategy planning and war fighting in the next 10 to 20 years will require highly sophisticated and well-integrated political-military-technical efforts. Strategy is now being rethought in the midst of what some believe to be a distinctive new revolution in military affairs.

As the SAIC team has pointed out, the current RMA is multifaceted and is demonstrating the simultaneous interplay and reinforcement of technical, operational, organizational, and socioeconomic developments. This "integrated system" RMA is also pushing the reinforcement of technical, organizational, and operational factors across all fighting mediums—air, land, and sea. Within this integrated-system RMA, some "new" or powerful areas of warfare are emerging—such as long-range and standoff precision strikes, information warfare, dominating maneuver emphasizing the strategic positioning of forces, and space warfare.

John Warden has written that offensive technical and military advancements, combined with suppression of enemy defenses, now give the United States the ability to wreak havoc on an adversary's entire target system. Such a degree of military shock might be applied so quickly that it could produce "paralysis." In time, however, adversaries may develop countermeasures to these US capabilities for parallel war and hyperwar, and they may discover how to reduce vulnerabilities of their operational, communications, and logistics centers of gravity.

Defensive measures are expected, and they usually can be countered; but if an adversary also is capable of using weapons of mass destruction, the stakes and the effects on strategy could shift substantially. WMD threats by radical, aggressive regimes may well force the United States into very different kinds of strategies. New modes of combat may have to emphasize mobile, indirect, dispersed, standoff, and disengaged operations until forces in the combat area can be defended against air and missile attacks.

The United States and its allies cannot risk putting large concentrations of troops and equipment in the way of a WMD attack. The magnitude of the casualties could exceed anything experienced by the United States in a single battle. For example, where the United States assembles a force like we fielded in the Gulf War or like we maintain in the Korea-Japan area, more US troops might be killed in a single one-day WMD attack than were lost in all the years of the Vietnam War, Korean War, or even in World War II.

In such megarisk situations, time-honored principles of war like "mass" may have to be reinterpreted to emphasize concentrating *firepower* rather than *troops*. An alternative is to rely more on constant movement, dispersion, outranging the enemy, deceiving the adversary about one's own centers of gravity, blinding enemy reconnaissance, and emphasizing disengaged "remote" combat until the enemy's WMD and other "big guns" are silenced. Unless the enemy WMD can be eliminated in initial counterforce strikes, maintaining the offensive initiative in combat may have to wait until we can erect missile and air defenses against them.

When confronting a "Saddam Hussein with WMD," it may become essential to develop and deploy an airtight air defense system and an effective multilayered missile defense in the regions threatened. Theater missile defense is the most important ingredient needed to cope with radical and well-armed regimes.

No less than a two-tiered defense system, where each layer has around a 90 percent probability of kill against an incoming enemy reentry vehicle, will be adequate to protect US overseas expeditionary forces, allied capitals, ports, air bases, naval convoys, and population centers. Anything less and the problems of dealing with NBC-armed adversaries begin to swamp the solutions. Without such protection, it could become suicidal to introduce an army into a port or put it into a region through local air bases. Absent effective missile and air defenses, WMD can scare off possible allies from joining a coalition against our enemy and raise the body count so high as to make US power projection into the region politically untenable. It could even threaten the outright defeat of US and allied forces in the field.

Without effective theater missile defenses, the costs of engaging such a NASTI may far exceed the gains in defeating him. Without effective missile defenses, it may even be advisable to revise US foreign policy commitments so we are not compelled to act against such lethal regional enemies.

On the other hand, the NASTI may be deterrable by allied superiority in WMD, or the early deployment of effective active defenses may help to persuade him not to escalate the conflict and to abstain from using his nuclear, biological, or chemical weapons. If such escalation dominance results in intrawar deterrence of the enemy's use of his worst weapons, then the United States and its allies may be free to exploit their technological edge via information warfare, precision strikes with advanced conventional weapons, space assets, and other strategies and techniques.

The development and employment of information warfare against an adversary's command and control systems, or against its leader's ability to appear legitimate in the eyes of its population, also looms as a potent new warfare technique. In 1991 information technology already was changing war fighting. The global positioning system allowed US and allied ground units attacking the Iraqi army's flank to maintain their positions accurately on the Kuwaiti desert even during blinding sandstorms. Self-navigating data drones can be employed to search autonomously across numerous information networks. Belligerents have already used propaganda via the Internet. Vulnerability to computer virus warfare and other nonlethal disabling technologies now has the attention of national security planners.

While the pursuit of nuclear weapons by rogue regimes is alarming, cheaper and quite lethal biological and chemical technologies are also being developed. The Iraqi biological warfare threat greatly concerned coalition military planners during the Gulf War. Saddam Hussein had large chemical weapons programs under way, and Iraq had begun to manufacture sizable quantities of BW agents prior to that war. As Mayer and Kadlec have written, biological warfare weapons are easy to produce, and it is estimated that at least two dozen countries have them. Some of the most dangerous among the BW resources are anthrax (bacillus anthracis) and the botulinum toxin.

As Mayer tells us, a cult in Japan spread the Sarin agent in Tokyo's subways, killing 12 Japanese citizens and harming

5,500 more. This group also had begun researching on and stockpiling biological weapons when Japanese police and security forces intervened. They unsuccessfully tried one biological weapons attack prior to their chemical weapons attack on the subway system. Biological warfare programs are hard to detect in the development and production stages, and biological agents could cause severe casualties if introduced into the water, air, or food supplies of crowded populations or unprotected armed forces.

Moreover, as Kadlec writes, biological agents can be easily adapted for use with commercially available sprayers, thus lending themselves to covert applications. Biological agents could be used to conduct economic warfare, and the agricultural disasters they cause may easily be disguised as natural events.

While the United States currently enjoys a significant conventional forces technical edge over likely rivals, this edge can be lost if a future adversary masters the tools of the last RMA or the next one. Present US and allied advantages will generate countermeasures by future enemies that neutralize or leapfrog them. There is little doubt that the continuous game of measure, countermeasure, counter-countermeasure, and so on, will continue into the twenty-first century as strategists and scientists introduce new technologies and applications.

Already, the United States and its allies are encountering rogue states newly armed with weapons of mass destruction that were formerly held only by the major powers. This will affect the ability to project power into those regions and will require a thorough reexamination of how future major regional conflicts are to be fought. Soon, too, the United States and its allies may encounter new modes of warfare—in the realms of land, sea, air, space, and cyberspace—through the use of innovations in computer-enhanced information technologies, digitized battlefields, space-based military systems, precision-guided weapons, theater missile defenses, and nuclear, chemical, and biological weapons.

Future enemies are unlikely to confront the United States in a straight force-on-force conventional battle. Rather, they might use asymmetrical strategies and techniques that utilize NBC weapons, guerrilla warfare techniques, state-sponsored terrorist attacks, satellite signal jamming, destruction of allied ground stations that receive information from space assets, and other information warfare techniques to level the playing field and achieve their aims.

Unfortunately, *Joint Vision 2010* gives few answers as to how the United States and its friends can cope with such strategies. More hard work and thinking are called for to inform the strategies and acquisition programs needed to cope with asymmetrical threats. National security is a continuous process. Those who rest on their present military advantages rather than seeking continuous improvements to cope with future threats and changing conditions will be left behind, consigned to defeat in the next era.

US and allied strategists, scientists, and operators must continue peering hard into the future to ensure mastery of these trends. This will help us stay ahead of diligent competitors and will give our military forces the highest probability of victory on the battlefields of the future.

Note

1. Andrew F. Krepinevich, "Cavalry to Computer: The Pattern of Military Revolutions," *The National Interest* no. 37 (fall 1994), 30–42. The 12 RMAs the author identifies since the fourteenth century include: (1) the infantry revolution in the Hundred Year's War where the six-foot longbow enabled archers to penetrate the armor of cavalrymen; (2) the artillery revolution which helped breach the walls of cities and castles; (3) the naval revolution when sails and cannons made oar-driven ships obsolete; (4) the sixteenth-century fortress revolution in which thicker walls and more intricate multilayered construction allowed static defenses to offset the earlier artillery revolution; (5) the Napoléonic revolution that harnessed the joint tools of the mass nationalistic army and the industrial revolution to a new strategy of independent marches and concentrations on the battlefield that hitherto had been impossible; (6) the land warfare revolution brought on by the introduction of railroad networks, the telegraph, improved range and accuracy of weapons due to rifling, and rapid-firing weapons of the late nineteenth century; (7) the naval revolution of the mid-to-late nineteenth century where wooden sailing ships with short-range cannons were replaced by steam-driven iron ships with long-range cannons, culminating in the dreadnoughts of the early twentieth century; and (8–10) the intrawar RMAs due to mechanization, aviation, and information transmission improvements. At the end of World War II, perhaps the most significant RMA of all began: (11) the nuclear revolution. This revolution in military affairs has recently been followed by (12) the information revolution based on digitization and computer applications.

About the Contributors

James Blackwell is the assistant director, Strategic Assessment Center, SAIC. He was previously the director of political military studies and senior fellow at the Center for Strategic and International Studies in Washington, D.C. A retired US Army officer, he earned his PhD at Tufts University, The Fletcher School of Law and Diplomacy, taught at West Point, and is the author of various publications including *Thunder in the Desert: The Strategy and Tactics of the Persian Gulf War* (Bantam).

Richard Blanchfield is a senior analyst, Strategic Assessment Center, SAIC. He specializes in research on the revolution in military affairs. In the past he served as a research analyst on the *Gulf War Air Power Survey.* A former US Marine Corps officer, Mr Blanchfield has 28 years' experience as an educator, administrator, pilot, operator, and planner, including 12 years of first-hand experience as an analyst working on political–military, national defense, and international relations issues. He served in his last military assignment as director, Studies and Analysis Division, USMC Warfighting Center. Much of his service career was focused on security issues in the Pacific or Middle East. A graduate of the US Naval War College (MA, national security and strategic studies), Mr Blanchfield also has an MA in International Relations from Salve Regina University.

Lawrence E. Grinter, professor of Asian studies, Air War College, teaches and writes on Asian and global security issues including WMD proliferation. He is the co–editor/author of numerous publications including *Security, Strategy and Policy Responses in the Pacific Rim* (Rienner), *East Asian Conflict Zones* (St. Martin's) and *Looking Back on the Vietnam War* (Greenwood), as well as 35 scholarly articles. Dr Grinter earned his PhD from the University of North Carolina at Chapel Hill.

Dale Hill is a senior analyst, Strategic Assessment Center, SAIC. His current work is focused on aerospace warfare in the study of the revolution in military affairs, where he has directed seminars, war games, simulations, and research. A former Air Force fighter pilot, Mr Hill commanded an F–16 squadron and served in senior positions as military assistant to the assistant secretary of the Air Force and as a military planner in Checkmate during Operations Desert Shield and Desert Storm. Mr Hill graduated from the Industrial College of the Armed Forces, the Marine Command and Staff College, and the Air War College. He has a Master's degree in international relations from Oklahoma State University.

Charles A. Horner, General, USAF, Retired, left the Air Force in 1995, after a 37–year career. His last position was as commander in chief, North American Aerospace Defense Command and the US Space Command. During Operations Desert Shield and Desert Storm, General Horner commanded all US and allied air operations. A command pilot, he has more than 5,300 hours in a variety of fighter aircraft, including F–16s, F–15s, and F–4s. General Horner was commissioned in 1958, and commanded wings, air divisions, the Ninth air force, and US Central Command Air Forces. General Horner flew over 130 combat missions in Indochina. He is a graduate of the University of Iowa, and has professional military education degrees from the Armed Forces Staff College, the Industrial College of the Armed Forces, and the National War College.

Robert Kadlec, Lt Col, USAF, is a physician assigned to the Office of the Secretary of Defense for International Security Policy as a senior assistant for counterproliferation. A specialist in biological warfare issues, he is a distinguished graduate of the US Air Force Academy. He holds a doctor of medicine and a master's degree of Tropical Medicine and Hygiene from the Uniformed Services University of the Health Sciences. He also holds a master's degree in national security studies from Georgetown University.

George Kraus is a senior analyst in the Strategic Assessment Center of SAIC. He has an extensive background in military and naval operational and intelligence analysis and of comparative analysis, technology assessment, and war gaming design. He specializes in information war research and has taught courses in his specialties at NDU and the Naval War

College. A retired US Navy commander, Mr Kraus worked as a military assistant to the director of net assessment, Office of the Secretary of Defense, and worked in Naval intelligence for many years. He has provided numerous research papers on current Russian security issues, submarine issues, command and control, information war, robot warfare, space military issues, and military issues relating to the former Soviet Union. A graduate of the Naval War College, Mr Kraus has an MA in international relations from Salve Regina College and an undergraduate degree in political science from MIT.

Fred Littlepage is deputy assistant, Strategic Assessment Center, SAIC, working on contract research on topics such as the revolution in military affairs. A former USAF officer, he has worked as the assistant director of research, Office of the Secretary of Defense, Net Assessment, served as assistant air attaché to the American Embassy in Moscow, was director of curriculum of the USAF Soviet Awareness Team, Directorate of Soviet Affairs, and served as a Minuteman combat crew commander. He is a Russian language and area specialist, and earned an MS in strategic intelligence at the Defense Intelligence College. He has published various articles that focus on the Soviet Union, the newly independent states, military tactics, and other intelligence matters.

Terry N. Mayer, Lt Col, USAF, is a master navigator with over 3,000 hours in HC–130, VC–135, and VC–137 aircraft. He served staff tours at every echelon from wing to air staff in various capacities. During the Gulf War, he was initial cadre on the Checkmate planning cell and later was drafted by DIA to lead a tiger team charged with finding a method of conducting a preemptive strike against Iraqi biological warfare storage and production facilities—a team that was awarded the National Intelligence Meritorious Unit Citation by the director of Central Intelligence. Most recently, Lt Col Mayer was the commander, 615th Air Mobility Support Group, Provisional, responsible for restructuring the concept of air mobility and global reach into the Pacific. He has an MBA from Chapman College and is a graduate of the Air War College.

Jeffrey McKitrick is the director, Strategic Assessment Center, for Science Applications International Corporation (SAIC) in McLean, Virginia. Previously, he served as the special assistant on national security and foreign policy issues to Vice President Dan Quayle. While on duty as an active duty US Army officer, he served as military advisor to the vice president; as the military assistant to the director of net assessments, Andrew Marshall; and as an assistant professor at the US Military Academy at West Point, where he taught courses in international relations, national security, and economics. He earned master's degrees at both the US Naval Academy and at the Johns Hopkins School of Advanced International Studies, after completing an undergraduate degree at Indiana University.

James W. McLendon, Colonel, USAF (MS, Troy State University) is a career intelligence officer. His experience includes national, ground tactical (mobile), and airborne operations. His overseas assignments include Taiwan (twice), Vietnam, Germany (twice), the United Kingdom, and Saudi Arabia. Stateside, he has served on the staff of Air Force Intelligence Command (AFIC) (and its predecessors) three times and was the AFIC Liaison Officer to the Air Force Special Operations Command prior to attending the Air War College. Colonel McLendon has commanded three intelligence squadrons, is a graduate of Squadron Officer School (residence), Air Command and Staff College (residence), and Air War College. He is assigned to Headquarters Air Force Special Operations Command as the director of intelligence.

Barry R. Schneider, professor of international relations, Air War College, writes and teaches a variety of issues concerning twenty-first century warfare, conflict and change, WMD proliferation issues, the revolution in military affairs, and international flash points. He has previously served as a college professor, defense analyst, foreign affairs officer (ACDA), congressional staffer, and active speaker and editor/writer on national security issues. He is the author and editor of many publications, including two books: *Missiles for the Nineties: ICBMs and Strategic Policy* (Westview) and *Current Issues in U.S. Defense Policy* (Praeger). Currently, he is finishing a book on "Counterproliferation: Military Responses to Proliferation Threats," Dr Schneider earned his PhD in International Relations from Columbia University.

George J. Stein, chairman, Department of Conflict and Change, Air War College, writes and teaches courses on information warfare, European security issues, and future strategies. Before joining the Air War College, Professor Stein was on the faculty of the School of Interdisciplinary Studies, Miami University and Miami's European Center in Luxembourg. He recently published "A Theory of Informaton Warfare: Preparing for 2020," in *Airpower Journal* (Spring 1995), and has received the Meritorious Civilian Service Award for his contribution to the USAF chief of staff's study *Spacecast 2020*. Dr Stein received his MA from Penn State and his PhD from Indiana University.

Richard Szafranski, Colonel, USAF, holds the National Military Strategy Chair at the Air War College. His BA is from Florida State University and his MA is from Central Michigan University. Colonel Szafranski's previous assignments include command at the squadron, group, and wing levels. Colonel Szafranski commanded the 7th Bomb Wing during 1991–1993, and was the commander of Peterson AFB, Colorado, in 1988. He has served at headquarters Strategic Air Command, the North American Aerospace Defense Command, US Space Command, and the Air Force Space Command. He is widely published in *Airpower Journal*, *Armed Forces Quarterly*, *Parameters*, *Strategic Review*, *Proceedings*, etc.

John A. Warden III, Colonel, USAF, Retired, completed 30 years of active duty in the United States Air Force in July 1995. He is a command pilot with over 3,000 hours in F–4s, F–15s, and OV–10s. A graduate of the Air Force Academy, the Air Force Institute of Technology, and the National War College, Colonel Warden flew 266 combat missions in Southeast Asia. He also played a leading role in planning and supporting the air campaign in the 1990–91 Gulf War. His book, *The Air Campaign*, is used in military schools around the world. Col Warden's last active duty assignment was as commandant, Air Command and Staff College.

GPO US GOVERNMENT PRINTING OFFICE : 2007—625-205